No. 1376
$13.95

ALL ABOUT CHIMNEYS– Building, Using, Maintaining & Repairing

by John E. Traister

TAB BOOKS Inc.
BLUE RIDGE SUMMIT, PA. 17214

FIRST EDITION

FIRST PRINTING

Printed in the United States of America

Library of Congress Cataloging in Publication Data

Traister, John E.
All about chimneys—building, using, maintaining & repairing.

Includes index.
1. Chimneys. I. Title
TH2281.T7 697'.8 81-18226
ISBN 0-8306-0055-8 AACR2
ISBN 0-8306-1376-5 (pbk.)

Contents

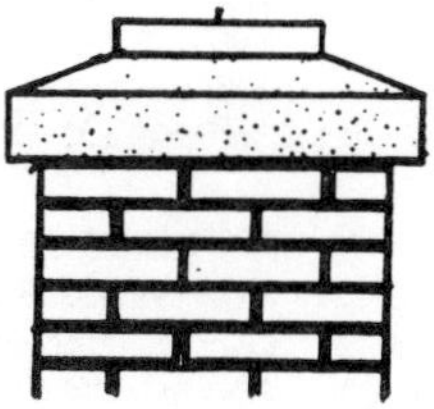

Introduction

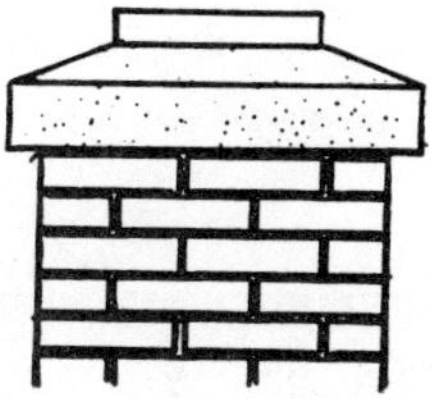

A quick look through any of the home and garden or popular science magazines reveals many ads offering woodburning stoves for sale. With increasing heating fuel costs, homeowners are turning to wood for supplemental heat. They are reducing their fuel bills substantially.

Many more homeowners would probably use wood heat if a flue was not required. Thousands of homes are without flues, particularly those that were built in the 1960s and use electric heat. Power companies at that time promoted electric heat which obviously didn't need a chimney or flue.

If you live in an old home whose chimney has not been used for years, the chimney probably needs repairing. This book shows you how to evaluate the condition of an existing chimney, make any necessary repairs, maintain your chimney properly, and even how to design and build new chimneys and fireplaces.

Other TAB books by the author:

No. 903 *Do-It-Yourselfer's Guide to Modern Energy-Efficient Heating and Cooling Systems*
No. 909 *How to Build Metal/Treasure Locators*
No. 923 *How to Build Your Own Boat From Scratch*
No. 925 *Practical Electrical Installation, Repair & Rewiring*
No. 996 *Treasure Hunter's Handbook*
No. 1018 *101 Vantastic Ideas to Improve Your Van*
No. 1020 *Automotive Air Conditioning Handbook—Installation, Maintenance and Repair*
No. 1126 *55 Easy Outdoor Projects for Do-It-Yourselfers*
No. 1131 *How to Tuneup & Repair Your Own Car*
No. 1137 *The First Book of Electronic Projects*
No. 1140 *Basic Gunsmithing*
No. 1298 *Professional Care & Finishing of Gun Metal*
No. 1317 *The Second Book of Electronic Projects*

Chapter 1
Chimney and Flue Inspection

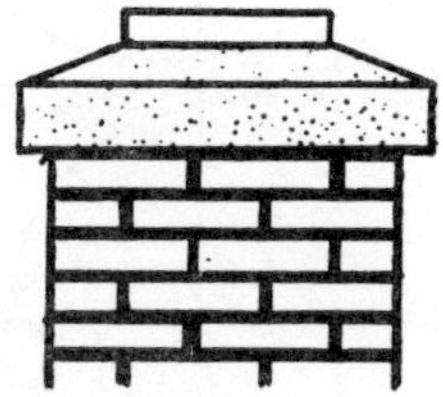

Perhaps your chimney has been idle for many years, and you now want to use it as a flue for a wood stove. Maybe you have been using your chimney to burn your wood stove. In either case, a survey of the chimney's condition is necessary to ensure proper operation and safety of your woodburning stove.

WOOD STOVE FLUE INSPECTION

Chimney flues for wood stoves should be ample in size and carried as nearly straight as possible from a point near the stove to above the highest projection of the roof. They should preferably be independent, having no connection with other flues (furnace, fireplaces, etc.) or openings, and always of the same area from top to bottom. A well-jointed tile flue with tightly cemented joints is a must—not only for proper operation, but for safety as well.

When looking over your existing chimney and connecting a woodburning stove to it, the following observations should be made.

- See that no openings exist into the flue, either above or below the stove's smoke pipe.
- It is best for the division walls of a chimney, if it contains more than one flue, to be carried up to the top of the chimney, and down to the bottom of the chimney, so that each flue is independent of the other throughout the entire length.
- The area of the chimney flue should be maintained full size throughout its entire length and be free from all obstructions such as loose brick and mortar that might become lodged in it. An offset in

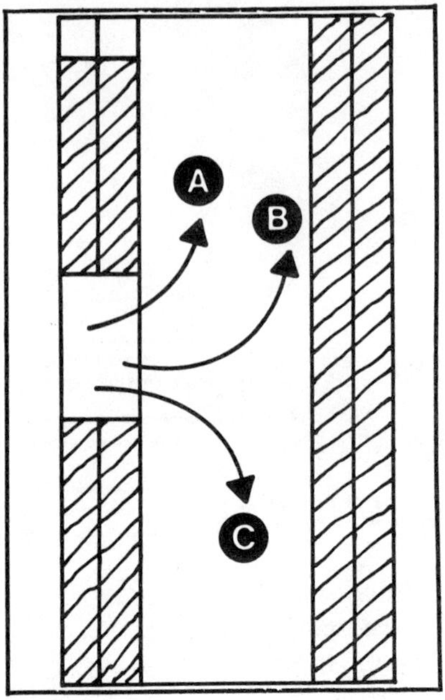

Fig. 1-1. The existence of another unwanted opening at the bottom of a flue, or a leak of any other type, may be readily detected by twisting a newspaper, and lighting and holding it against the opening where the smoke pipe enters the chimney flue. If part of the flame goes downward (C) this indicates a leak below the thimble opening.

the flue should have increased area to overcome the added friction of the offset.

■ Look for an oily substance like *creosote* on the inside of the flue. Creosote is usually caused by burning green wood in a tightly closed stove. This substance can ignite and is often the cause of chimney fires. A mirror may sometimes be used to detect obstructions and the general condition of the flue.

■ The chimney should extend above the highest point of the roof or other immediate surroundings such as adjoining buildings, hills, or trees. If not, you can expect a poor draft.

■ The smoke pipe should not project too far into the chimney.

To test for other openings or leaks in a chimney, twist a newspaper and light the end. Then hold the lighted newspaper against the opening where the stove's smoke pipe will enter. If part of the flame goes downward, there is a leakage below the smoke pipe entrance (Fig. 1-1).

In most cases, when the test indicates a leakage below the opening, it is caused by two flues being joined together at the bottom for cleanout purposes—using only one cleanout door for two or more flues. New cleanout doors should be provided at the bottom of each flue (as described in Chapter 2). The existing or middle cleanout door is used to seal off the two flues with cement.

Where it is necessary to have thé flue extend below where the stove's smokestack or pipe enters, unsatisfactory results may be avoided by giving the end of the smoke pipe the form indicated in Fig. 1-2. Care, however, must be taken to see that the pipe does not become turned the wrong way.

WOOD STOVE INSPECTION

While you're checking your chimney, you'll also want to give your wood stove a check; this should be done yearly. Wood stoves listed by Underwriters Laboratories (UL) or approved according to UL standards come from the manufacturer with instructions on ideal use and maintenance. Most of these approved woodburning stoves are airtight and can be closed off in the event of a chimney fire.

Creosote, the main ingredient in flue buildup, comes from the burning of soft or green wood. Pine is the most notorious and should be avoided as much as possible except for kindling. Another mistake many people make in using wood stoves is purchasing a wood stove larger than needed to heat a particular area. For example, a stove designed to heat a 2,000-square-foot area operating in a 1,500-

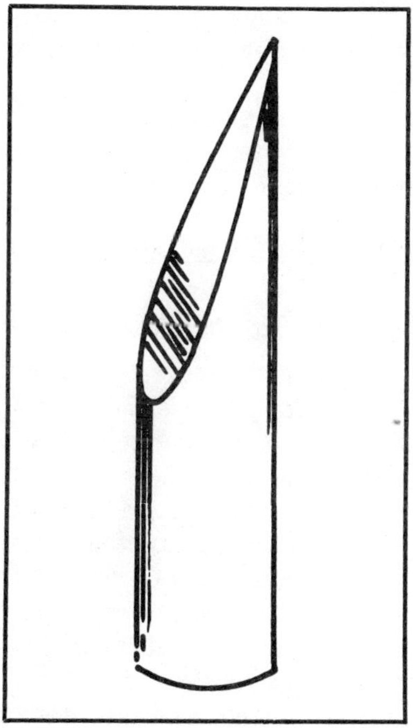

Fig. 1-2. Method of cutting stovepipe for better results when the flue extends below where the pipe enters the chimney.

Fig. 1-3. Dangerous creosote and soot buildup in stove pipes can prevent your chimney from drawing properly. This is also the chief cause of chimney fires.

square-foot area will burn at a lower rate than required. There is more chance of creosote buildup (Fig. 1-3).

The *Virginia Uniform Statewide Building Code* recommends the following suggestions when installing woodburning stoves.

- Be sure there is an 18-inch clearance between the stove and floor with a 24-gauge steel pad placed beneath the appliance.
- Check the stovepipe regularly for corrosion and holes. Replace if necessary.
- Install the small end of the stovepipe down, or pointed toward the stove.
- Avoid long horizontal runs, and be sure the pipe is securely supported.
- Minimize the number of elbows.

REMOVING AND CLEANING STOVEPIPE

One of the biggest problems with the installation of a new woodburning stove is the *thimble* area of the chimney—the hole where the stovepipe enters. If the wall is paneled or contains wood framing members, there should be at least 8 inches of masonry between the thimble and the wood or other combustible members (Fig. 1-4). This situation, if left unchecked, causes the thimble to heat up and possibly to ignite the wood.

In the case of a woodburning stove or freestanding fireplace, first remove the metal stovepipe between the unit and the thimble in the chimney. There will be some falling soot from the pipe, so be sure to provide adequate protection for the carpet or floor. A

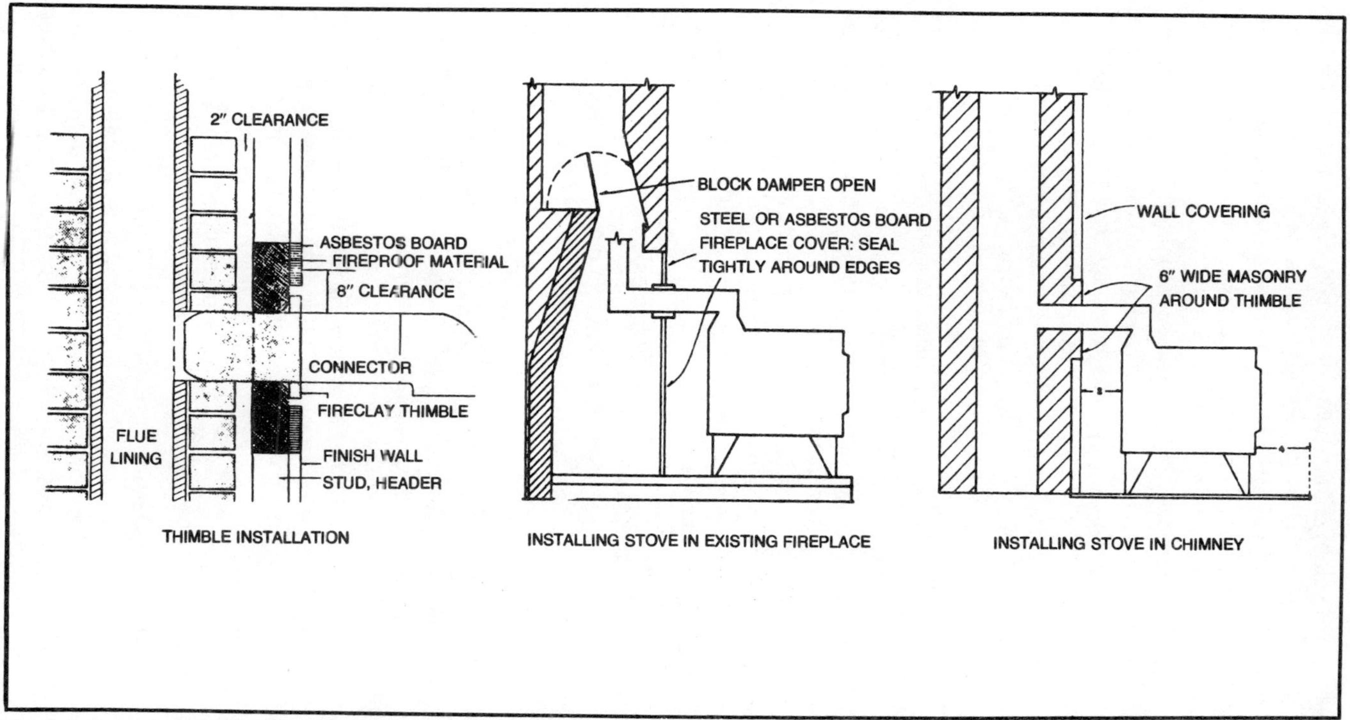

Fig. 1-4. Correct methods of installing several types of woodburning stoves.

painter's drop cloth is fine. If you don't have access to one, merely lay down several layers of newspapers. Have a vacuum cleaner handy—preferably a heavy-duty industrial wet/dry type. Besides floor protection, you'll also want to protect the wall area (where applicable) from the thimble down to the floor.

Carefully remove the stovepipe with the least amount of knocking as possible to try to retain a lot of soot in the pipe. The pipe in large sections, should be taken outside for examination under sunlight. Examine each section thoroughly, especially around the joints, as this is where the pipe will begin to deteriorate first. Any worn or defective pipe should be replaced. Black pipe is the least expensive, but galvanized pipe will last the longest. Many homeowners prefer black pipe as it gives a more traditional look to their homes.

When the defective sections of stovepipe have been replaced, or if the existing pipe is okay, give all sections a thorough cleaning. Most chimney fires start in the stovepipe between the woodburning stove and chimney, as this is where most soot and creosote collects. The roaring blaze in the stovepipe can easily trigger a fire within the chimney. Since the heat of this pipe can soar to the melting point (around 3000°F), combustibles within the area are easily ignited. Even if you don't have your chimney cleaned very often, you can usually prevent a chimney fire by keeping the section between the woodburning unit and the chimney thimble clean, especially if the distance is greater than 4 feet. Experts recommend that a thorough cleaning of the stovepipe every two or three weeks will almost guarantee against a chimney fire. Even if your chimney is dirty, flames from the stove seldom rise or travel into the chimney itself; they are usually confined to the stovepipe.

While you have the stovepipe out, you might consider a damper in the pipe above the stove if you don't already have one. Pipe sections are available in conventional sizes with the damper built in. This will let you have better control of the fire in your woodburning unit and will also help guard against the flames from your firebox reaching the inside of your chimney.

CHIMNEY INSPECTION

If your chimney is for a conventional fireplace (with no metal stovepipe connection), completely clean the fireplace. Remove all ashes. Clean all loose soot from the walls of the fireplace, the ceiling, and around the damper door. You don't have to be too particular as you'll have to clean it again after your inspection is finished. Soot and ashes are going to fall from the chimney flue during the inspection and cleaning.

For a masonry chimney to which a stovepipe is attached, use your vacuum cleaner and remove all soot from the outside of the thimble opening to where it attached to the chimney flue. A wire brush may be necessary to remove some of the soot and other debris. Again, make sure your wall is protected directly under the thimble opening. Newspaper attached to the wall with masking tape will normally do the trick.

Flue

A mirror and flashlight will allow you to examine your chimney flue. One of the small mirrors like those in ladies' compacts is ideal. Attach this to a small slat to project it back into the chimney through the thimble opening, and then use a flashlight as shown in Fig. 1-5. You should have a full view of your chimney from the thimble opening to the top. If enough light comes in through the top of your chimney, the flashlight may not be necessary.

You should have a pencil and pad handy to take notes as you view every inch of your chimney flue. I use a small cassette tape recorder. When I want to note points of interest, I merely talk into the tape recorder, which is continuously running, and my message is recorded for later reference. This mirror/flashlight inspection will allow you to detect animal and bird nests that must be removed before the chimney can be operated safely and effectively. Note the condition of the flue lining and the amount of soot and creosote present, the condition of the flue joints, etc. This inspection will also let you know if your chimney is lined with flue liner or merely built out of bricks. If the latter situation is the case, you may want to make plans to have a flue lining installed.

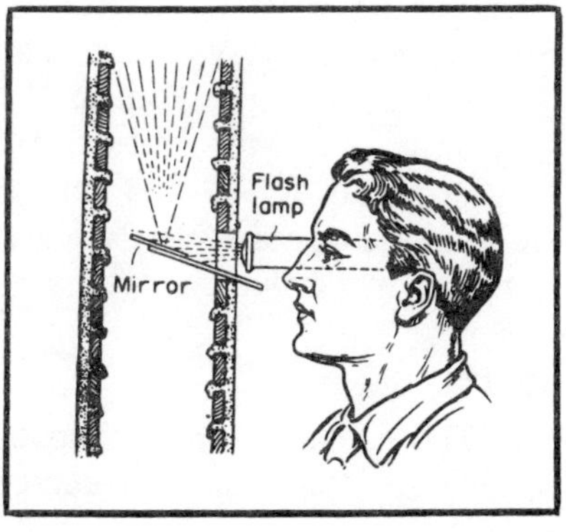

Fig. 1-5. Use a mirror and flashlight to view the inside of your chimney, looking through the thimble opening.

Fig. 1-6. Examination of this chimney reveals loose mortar and bricks. A closer look would probably show that the flashing around the chimney is in need of repair.

Once you have completed the preliminary inspection of your chimney, you must then determine how to go about correcting or repairing the problems. After reading the chapters on chimney repairs and cleaning, you may decide to have all the work done by a professional. You may want to do the work yourself. If there are several nests and the creosote buildup has reached a dangerous point, you may be better off hiring a professional chimney sweep. He'll remove all nests, clean the walls of the chimney, and free them from soot and dangerous cresote.

Joints

The outside of your chimney comes next. Usually a quick visual inspection from the ground will reveal the condition of your

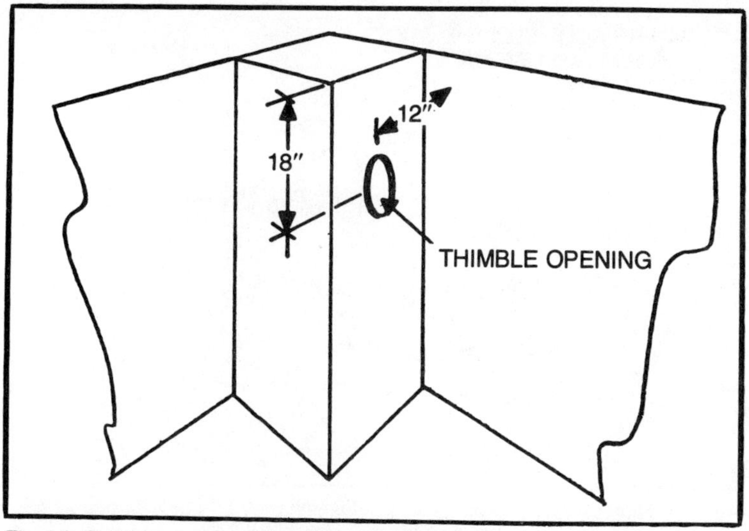

Fig. 1-7. Thimble openings should be at least 18 inches below the ceiling and 12 inches away from any combustible wood members.

overall chimney. If the masonry joints look tight, chances are all of them will be tight. If some of the joints at ground level to eye level are deteriorating, you'll probably have the same thing happening farther up. Any loose bricks should be repaired (Fig. 1-6). If some of the joints seem in poor shape, the entire brick joint should be scraped out. New mortar should be applied to insure a tight fit.

Should several of the joints show deterioration, you should use a ladder and inspect all joints carefully. Wait until dark, have a

Fig. 1-8. Step flashing is suitable for most roofs.

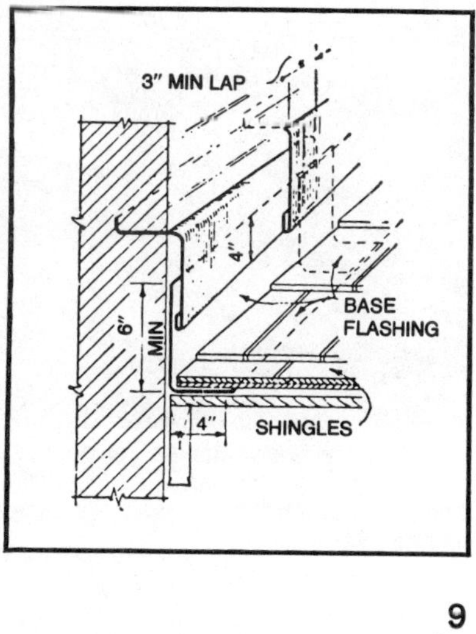

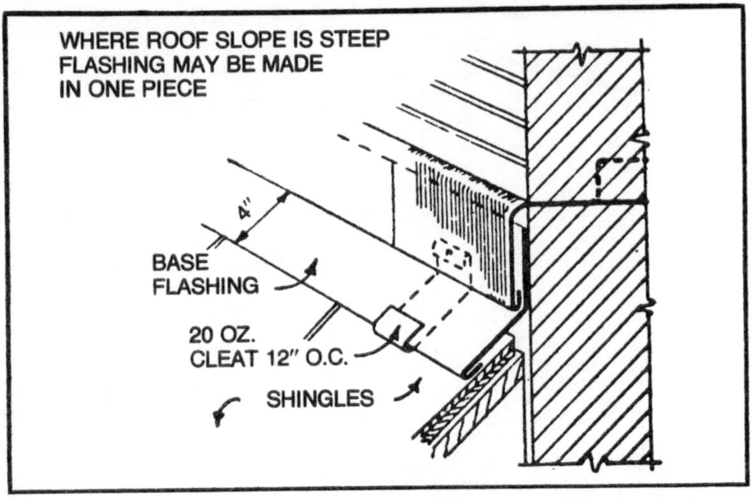

Fig. 1-9. Where the roof slope is steep, flashing may be made of one piece.

helper shine a relatively powerful light up or down the chimney flue, and then carefully inspect each joint to make certain you can't see any light through the joints. If so, then the joint should be repaired. You might want to carry a piece of chalk with you as you make the inspection to mark the various defective areas. Where the chimney runs through an attic or other spaces within the house, you should inspect these places the same way.

Fig. 1-10. The water stained ceiling tiles and damp insulation obviously mean there is a leak somewhere. The inspection showed that the flashing around the chimney was defective. Water was seeping through to run down the chimney and ruin the ceiling.

Fig. 1-11. Cracked plaster around a chimney is a sure indication that moisture is coming from somewhere. Check the chimney crown and flashing around chimney.

Leaks

Another consideration during the inspection is the distance chimneys and chimney openings (thimbles) are from wooden struc-

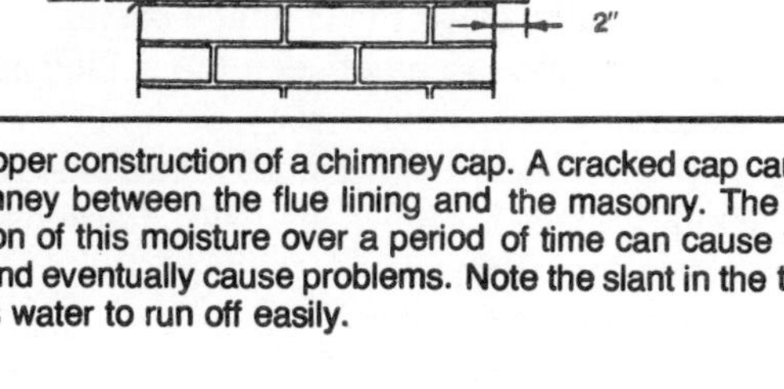

Fig. 1-12. Proper construction of a chimney cap. A cracked cap can let moisture into the chimney between the flue lining and the masonry. The freezing and thawing action of this moisture over a period of time can cause the mortar to deteriorate and eventually cause problems. Note the slant in the top of the cap which allows water to run off easily.

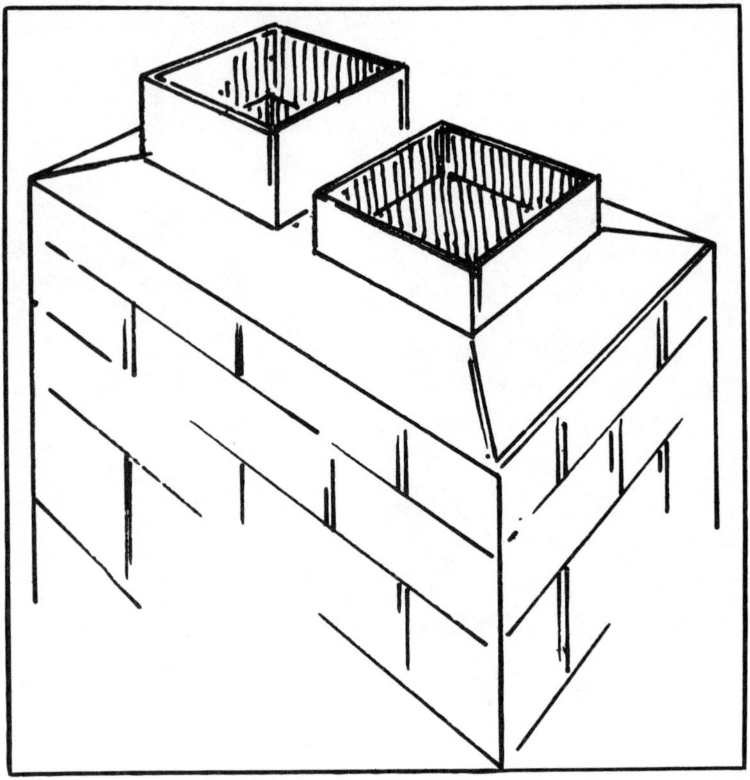

Fig. 1-13. When two flues are in the same chimney, one should be higher than the other to prevent the draft from one flowing down the other.

tural members (Fig. 1-7). A roaring chimney fire can reach temperatures in excess of 3000°F. Framing members within 12 inches or less of an opening (thimble, defective joint, etc.) can ignite, causing an entire home to burn. One place to be especially cautious is where a chimney thimble opening is located near a combustible ceiling. A chimney fire can melt the metal stovepipe or cause it to get hot enough to ignite a combustible ceiling that's too close to the opening. Such conditions can be corrected by placing a sheet of noncombustible material directly above the chimney opening—preferably on spacers to allow an air space between the ceiling and the noncombustible material.

If you have a clean flue, free of animal or bird nests and little or no creosote buildup, many homeowners have checked for leaks by building a small fire directly inside the chimney. Build a small fire of wood and tar paper at the base of the flue or in the fireplace. When smoke begins to pour heavily from the chimney top, cover the top with a heavy wet burlap bag or an old blanket. Inspect the chimney carefully to see if you can find any smoke escaping through the masonry. If you do find any leaks (by this or any other method), the cause must be corrected before the chimney is safe.

The thimble opening should provide a relatively tight fit for a stovepipe—tight enough so that drafts will not cause smoke to flow out the sides and fill a room. Furthermore, the pipe should be secured tightly enough so it does not come loose due to vibrations or an accidental jarring of the stove. Do not, however, have the opening too small as the pipe will crimp when it is fitted into the opening. This will obstruct the flow of smoke and hot gases and may result in creosote buildup.

Many masonry chimneys are built flush with the inside wall and then plastered over the brickwork to match the adjoining walls. After a house settles, there will be an obvious crack between the chimney and the adjoining walls. Seal this crack.

You'll also find that chimneys (especially those that are built on the interior walls of a house) are trouble spots for leaking around the chimney flashing (Figs. 1-8 and 1-9). This leaking will rot wooden

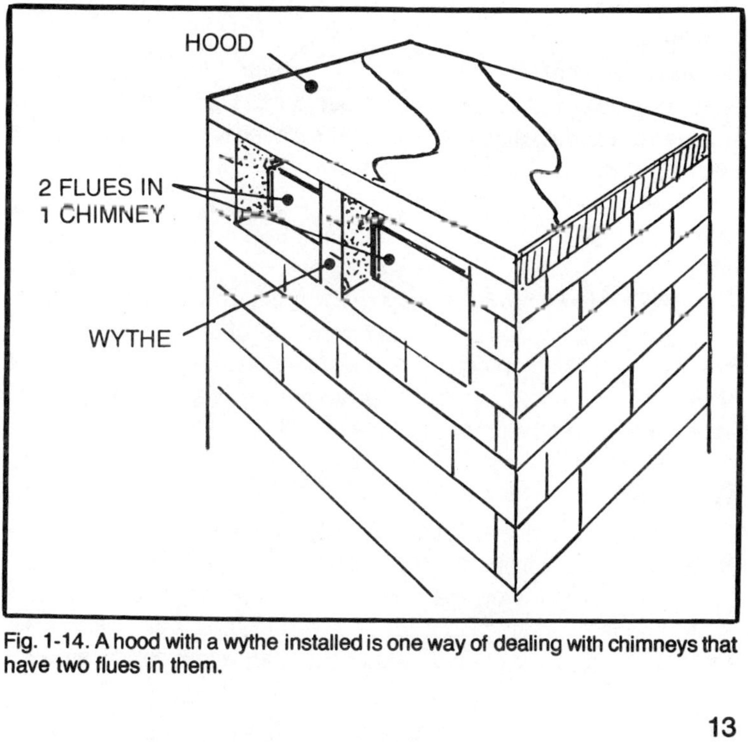

Fig. 1-14. A hood with a wythe installed is one way of dealing with chimneys that have two flues in them.

Fig. 1-15. While this chimney is slightly higher than the roof ridge, it would draw better if it were about 10 inches higher.

members and spoil ceiling plaster and the wall plaster in the immediate area. When making your chimney repairs, check for leaks around them (Figs. 1-10 and 1-11). Much repair work on your home can be saved if you take care of these leaks immediately.

Footing

Did you consider your chimney *footing*? Most chimneys and fireplaces have footings at least 2 feet deep. Your chimney footing will be the least of your worries, but it's a good idea to check it when you're inspecting the chimney. If repairs are necessary, I'd recommend getting professional help.

Cap

The chimney *cap* is usually made of poured concrete to keep moisture, dust, and other foreign matter out of the masonry. Chimney caps will develop cracks that allow moisture to seep into the seams of the masonry work (Fig. 1-12). The joints may be weakened. Inspect the cap and make necessary repairs.

Hood

Consider chimney accessories such as a *hood* to prevent gusts of wind or drafts from blowing down the chimney. When considering such a device, however, make certain that the cross sectional area of the flue is maintained. The total area left open must be at least equal to the cross sectional area of the flue. Some builders prefer just a trifle more. You might choose to close the hood on two sides and open it on the sides to the prevailing winds. This promotes a sustained draft while eliminating downdrafts.

Consider the effect of two flues in the same chimney during your inspection. One should be higher than the other to prevent the draft from one flowing down the other (Fig. 1-13). You may also correct any draft short circuiting by installing a hood and then separate the two flues with a *wythe* (Fig. 1-14).

This just about covers your chimney from top to bottom. If any things crop up during your inspection, correct them. Just use common sense and the suggestions covered in this chapter, and you should be in good shape as far as evaluating the condition of your chimney (Figs. 1-15 and 1-16).

FIREPLACE INSPECTION

If you have a fireplace and a woodburning unit installed within the fireplace, or a woodburning stove connected to your fireplace flue and the fireplace opening sealed off, you should first remove the woodburning unit before the inspection begins. Clean all parts of the woodburning unit and stovepipes.

You can begin your inspection by checking the fireplace footing. Since the chimney is the heaviest part of most homes, it must be built on a solid foundation of concrete. The minimum size should be 12 inches thick and project 6 inches beyond the chimney. If your

Fig. 1-16. This chimney has been neglected so long that bricks have fallen away from the chimney tile. A complete renovation is necessary, not to mention the ruined interior finishes.

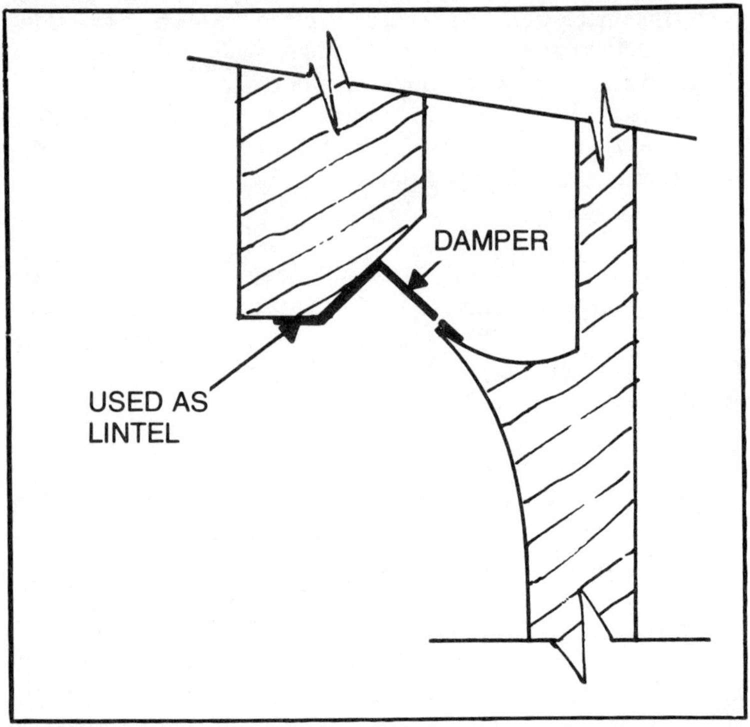

Fig. 1-17. Some dampers are used as lintels which can result in smoking fireplaces.

foundation is not of adequate size or in need of repair, seek professional advice.

One part of a fireplace that is often overlooked is the *ashpit.* Although not absolutely necessary, it is a great convenience for removing ashes without getting them all over the living room floor. The greatest faults include a rough passageway to the cleanout door, which promotes ashes clogging on their way downward. The worst problem is allowing water to leak into this space. When water is mixed with ashes, a cementlike mass is formed that makes the ashes extremely difficult, if not impossible, to remove.

In inspecting your ashpit, make sure the walls and floor (just below the cleanout door) are smooth. Then make absolutely certain that the area is watertight. This means no leaks in the masonry and a gasket on the cleanout door, especially if it's located on the outside of your home and is exposed to the elements. Make sure the back of your fireplace is sloped to reflect heat outward and to direct smoke and hot air toward the throat of the fireplace.

See that your damper is functioning properly. The size of the opening is controlled by raising the valve plate with a lever or screw. A damper door that will not close completely will suck enormous amounts of hot air up the chimney when you're using some other type of heat.

The damper is another part of the fireplace that is often installed incorrectly. The lintel across the front of the fireplace should be separate from the damper frame. Sometimes the damper frame is used as a lintel (Fig. 1-17). If this is the case, the easiest solution is to lower the front opening of the fireplace by adding a hood (Figs. 1-18 and 1-19).

Another fault is the damper is sometimes shorter than the width of the fireplace opening; it should be the same width. You can correct this error in two ways: replace the damper with a longer one

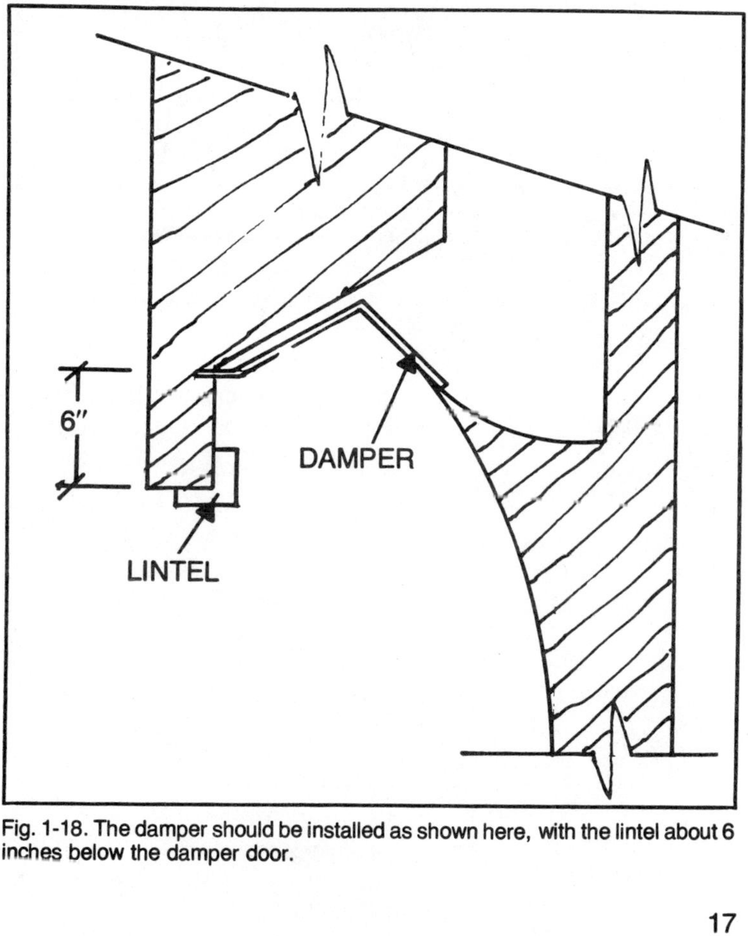

Fig. 1-18. The damper should be installed as shown here, with the lintel about 6 inches below the damper door.

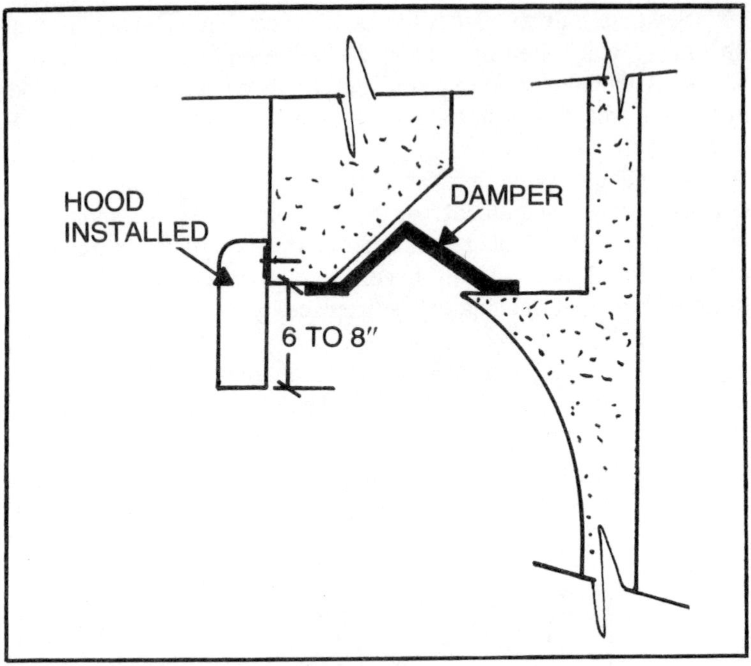

Fig. 1-19. One way to correct a faulty damper door is to install a hood across the firebox opening to make it 6 to 8 inches below the damper door.

or make the firebox narrower by building up the inside with firebricks.

The fireplace should have a smoke shelf to keep cold air current from flowing down the chimney. You can help a defective smoke shelf somewhat by installing a hood on the top of your chimney.

Chapter 2
Chimney Repairs

You will more than likely need to perform some necessary repairs to insure an efficient chimney. This chapter is designed to instruct you on the most common repairs to put your chimney in first class condition.

A good, well-maintained chimney is extremely important for satisfactory operation of any woodburning appliance. This is one reason why local building and fire codes apply to all chimney installations.

A warm chimney is more desirable than a cold one. Although some creosote will always be formed, the warmer the chimney, the less problem you are likely to have with creosote and acid formation. A warm chimney also increases draft. Installation of the chimney in the home's interior will aid in maintaining a warm chimney and will radiate additional heat to the living area (Fig. 2-1).

Downdrafts are frequent problems with chimneys (Fig. 2-2). To minimize downdrafts, the top of the chimney should be at least 3 feet above flat roofs and 2 feet above a roof ridge or raised part of the roof within 10 feet of the chimney. Under certain wind conditions, tall buildings or trees near the chimney top may still create downdrafts. In such cases, increasing the height of the chimney, or adding a booster fan or turbine at the top, may be a solution.

Approved metal chimney kits are efficient and easy to install. Metal chimneys must be (UL) listed for each type of stove used. Clearances should be marked on each section of the chimney. The factory-built chimney may pass through a floor if provided with a continuous fire-resistant enclosure approved for the purpose. Avoid using single wall pipe installed through a window, wall, or roof. The

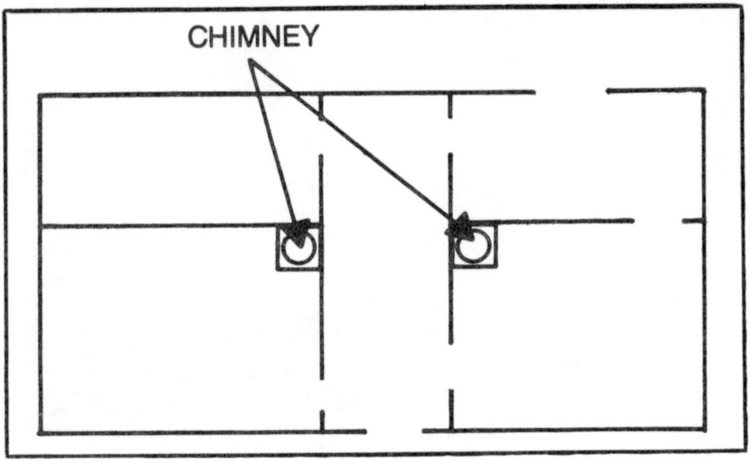

Fig. 2-1. Installing a chimney in the interior of the house will aid in maintaining a warm chimney and will radiate additional heat to the living area.

pipe will cause serious creosote, acid condensation, and corrosion problems. In addition, single pipe used as a chimney will stay colder, thereby reducing draft.

An existing chimney may be used only if it is sound. If the chimney lacks a flue liner, or if the masonry or lining is cracked, consult a mason about repairs. Techniques are available for installing a flue liner in an existing chimney. Fire in an unlined or cracked chimney can permit sparks or burning creosote deposits to enter the house, resulting in disaster. If you have doubts about your chimney's safety, consult a contractor, fire safety inspector, and your fire insurance company.

Two or more stoves connected to one chimney may not draw properly or comply with local fire and building codes. For best results, avoid connecting more than one stove to a chimney flue. In no case should a flue serving a fireplace or open-front Franklin stove serve any other heating unit.

REMOVING SOOT AND CREOSOTE BUILDUP

A major problem with woodburning is possible buildup of creosote and soot in the stovepipe and chimney. On new or clean burning stoves, a tap on the stovepipe with your index finger will result in a "ting" sound. When that sound changes to a muffled "thud," there is probably a substantial amount of creosote buildup in the pipe and chimney. This may result in a dangerous chimney fire unless cleaned. If you intend to burn wood continuously, some time will probably be necessary each year for chimney maintenance. Chemical chimney cleaners are not particularly effective in removing heavy deposits. They should be used with caution.

A more effective approach may be to seal the front of the fireplace and keep soot and dirt out of the room. Use one of the mechanical methods of cleaning. A rope may be used to lower and raise tire chains inside the flue, or a burlap bag filled with straw or chicken wire may be pulled up and down the chimney flue. Another method of cleaning the chimney is to work a small evergreen down it. When the tree is pulled back up, some of the creosote and soot will be dislodged. Commercially available flue cleaning brushes are also available (see Chapter 3). When cleaning the chimney, be careful not to knock out mortar joints or damage the flue lining.

Chimney fires are usually started by roaring blazes which send a mass of burning materials up the flue. Burning of paper, twigs, or other light material may also trigger a chimney fire. If, in spite of good housekeeping and precautionary measures, you should have a chimney fire, call the fire department. An airtight stove is an advantage since the fire may be controlled by closing the damper on the stove to reduce oxygen supply. In order for this to work, the damper must be solid and stovepipe connections must be tight. A sound chimney with no other stovepipe connections is also needed in this case, which is one reason why one stove connection per flue is wise. While awaiting the arrival of fire fighters, the *National Fire Prevention Association* recommends that you control a chimney fire by dumping large quantities of rock salt or baking soda into the fire chamber of the stove. Do not, however, put water in a woodburning

Fig. 2-2. A chimney hood for this 8-inch stovepipe helped to prevent downdrafts as well as to keep out rain.

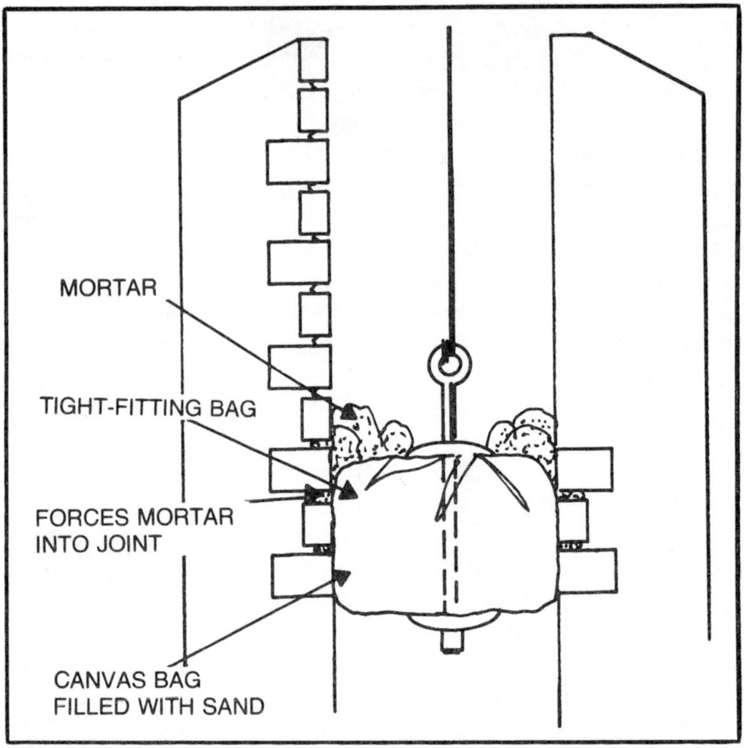

Fig. 2-3. A canvas bag filled with sand will force mortar into loose joints inside of the chimney.

stove or into the chimney itself because of the hazard of steam explosion, or the possibility of cracking the stove firebox or chimney flue liner.

MASONRY REPAIRS

If your inspection has detected cracked cemented joints or if the bricks are flaking or broken, they should be repaired as soon as possible to prevent further deterioration. Any loose cement should be cleaned out with a masonry chisel. A stiff mortar should be mixed consisting of 1 part mortar cement to 2 parts sand. This mixture is then forced into the joints. On the outside of the chimney, the job is no real problem as conventional masonry techniques may be used. On the inside of the chimney, it is a different story. One of the easiest ways to repair mortar joints inside a chimney is to fill a canvas bag with sand (Fig. 2-3). A rod is run through the bag and secured with an eye on the top end to affix a rope or chain. The bag should fit tightly, but be free enough to slide up and down the inside of the chimney.

Mortar is then placed on top of this bag, after it has been partially placed into the chimney, and lowered down the chimney. At each joint that needs to be repaired, the bag is moved up and down—above and below the joint—to allow mortar to seep over the side of the canvas bag, and then to be forced into the joints.The technique is a little tricky, but with a little practice you should be able to get enough of the mortar forced into the joints to seal them. Be sure the bag fits the chimney liner properly and that the mortar mix is of the right consistency. Then it's merely a job of pulling the bag up and down the chimney liner, and keeping enough mortar on the top and around the sides of the bag.

The most likely place for your chimney to need repairs is the chimney cap. If the cap is cracked or chipped, it can allow water to leak down between the flue and brickwork and, by freezing and thawing, this action can cause serious damage to the masonry work. If the chimney cap is in bad shape, it is probably best to knock the entire cap loose and start from scratch. You can then build a form from wooden boards and pour a new cap—using reinforcing rods in the form to help hold the cap into shape and prevent it from cracking again in the future. In finishing the cap, make certain that it slopes away from the flue liner so that all water will run off *away* from the chimney opening (Fig. 2-4).

In new chimney construction, or when you're rebuilding an existing one, allow the flue tile to extend about 4 inches above the last course of bricks or concrete blocks. Build a wood form around the top of the chimney brickwork, and pour a 3-inch cap of concrete

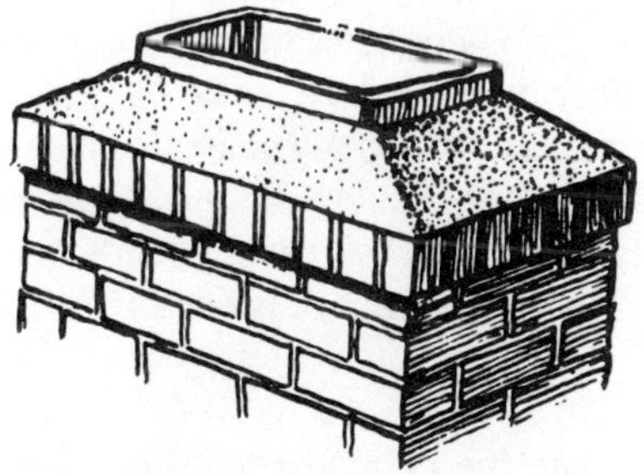

Fig. 2-4. Chimney cap must be free from cracks and be slanted to let water run off. Water getting in between the flue liner and the masonry can cause the joints to deteriorate.

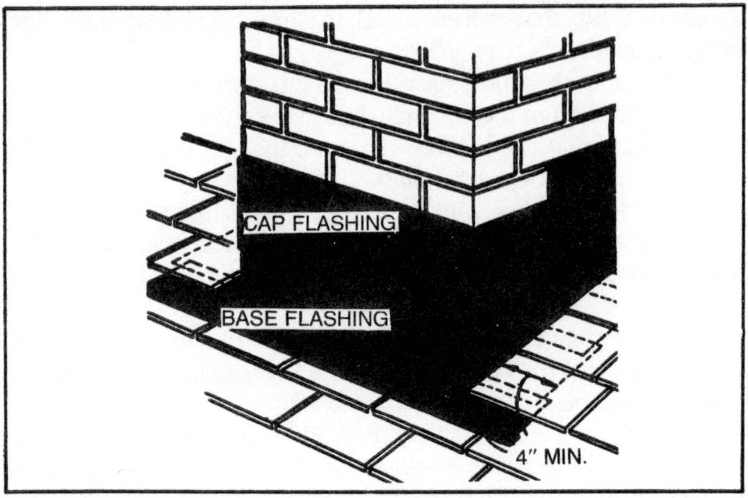

Fig. 2-5. Flashing is normally installed in two sections, the cap flashing and the base flashing.

over the bricks. Give the cap a slight downward slope so that water will drain off easily.

If an existing cap is of the proper design but cracked, clean out all cracked joints. Point the cracks with an approved waterproof *mastic compound.*

You may wish to clean the masonry work both inside and outside of your chimney and fireplace. The safest way is to try scrubbing the brickwork with a strong detergent and water. If this doesn't do the job, you can try a mild acetic acid diluted with water according to the manufacturer's instructions. For real tough stains, scrub with a solution of *muriatic acid* (1 part acid poured slowly into 10 parts water, not the other way around) and then rinse off immediately with water. When using muriatic acid, be sure to wear eye protection (safety goggles, face mask, etc.) and rubber gloves. Its use is only recommended for the seasoned mason. You may try using certain abrasive compounds helped along with a stiff wire brush for the cleaning job. These compounds are also good for removing soot stains on fireplace interiors.

FLASHING

The joint between the chimney and the roof must be flashed with metal to make it watertight. The flashing consists of the cap flashing and the base flashing (Fig. 2-5). The cap flashing should be bent at a right angle so that it can be fastened right into the mortar joints. It should extend in about 1½ inches. All joints should overlap 4 inches and be made tight with roofing or mastic compound. Note in Fig. 2-5 that the roofing or shingles overlap the base flashing by at least 4 inches.

Flashing may consist of any pliable metal, but copper is preferred and tin is the second choice. Most modern repairs are done with galvanized sheet metal since copper is so expensive.

A metal "cricket" in back of the chimney is desirable on certain roofs so that water running down the roof will not collect and seep through any openings (Fig. 2-6). The dimensions will vary according to the size chimney, roof slant, and the like. Note the L-shaped flashing and the arrangement of the counter flashing in relationship to the chimney. A metal cricket or saddle with soldered joints is the preferred method.

It may be desirable to cut a new opening in an existing chimney to allow a stovepipe to enter. The bottom of this opening should be 8 inches from the bottom of the flue tile. Fill a bag with damp sand and stuff it down into the flue at the point where the opening is to be cut; then tap a cold chisel very lightly with a hammer along the line where the cut is to be made. You can purchase special thimble sections of flue tile that already contain the opening for the stovepipe. Before you cut the opening or purchase the special thimble section, check the size of the stovepipe you plan to use. An 8-inch diameter is standard for most woodburning stoves.

The joint between the thimble and the flue tile and the surrounding masonry must be tight at all points. Furthermore, the opening must be at least 18 inches below the ceiling. It should be at least 12 inches away from any woodwork or combustible materials. The thimble must be set absolutely level so that the metal stovepipe can enter it straight.

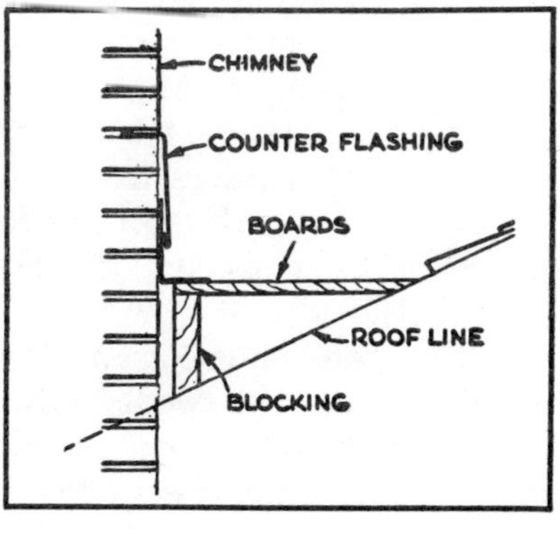

Fig. 2-6. A cricket is sometimes used to prevent snow and water from collecting in the gulley between the chimney and the slanting roof line.

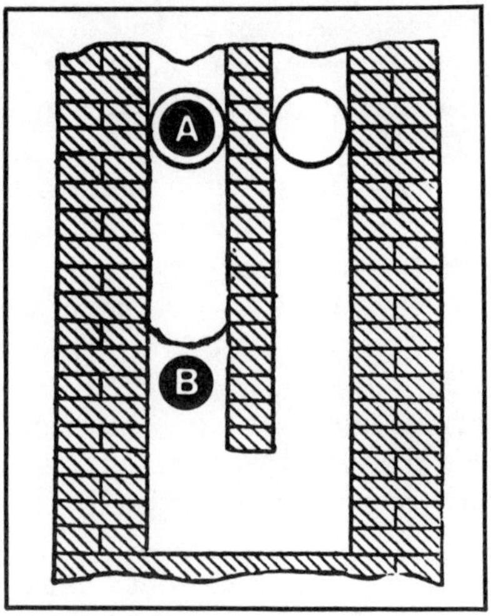

Fig. 2-7. A metal plate tightly fitted at point B will act to separate two flues in one chimney when they are not independent of the other.

CLEANOUT DOORS

It is also desirable to have a cleanout door at the base of the chimney. This is not absolutely necessary, but it certainly makes cleaning the chimney much easier. The flue tile lining is carried down past the opening for the stovepipe. The top of the door should come almost flush with the bottom of the first section of flue tile. Metal doors for cleanouts can be purchased from furnace and fireplace suppliers.

Cracks in plastered chimneys inside the home may be repaired with a plaster mix or covered with drywall. The seams are spackled in the conventional manner. When repairing cracks, the crack should first be cleaned and gouged to allow an area for the repair mixture to adhere.

Chimneys that are built too close to wooden members must be corrected before they are considered safe. This will often require extensive carpentry work. You might want to consult an expert. Where the chimney runs up through a ceiling or roof and the joists or rafters are too close to the masonry, the openings will have to be reframed. The members that are too close to the chimney are then removed. A mistake at this point can result in serious damage to the structural design of your home. If there is any doubt, seek professional help.

The condition in Fig. 2-7 is very commonly encountered in older chimneys with two flues. Note that the chimneys are joined together at the bottom and will cause a short circuit between them interfering with proper drawing of either flue. This situation can be remedied by inserting a sheet of iron in the flue tightly at the point indicated in Fig. 2-7B.

A similar situation occurs when one cleanout door is used for two flues (Fig. 2-8). The division walls of the chimney should be carried up to the top of the chimney and down to the bottom, so that each flue is independent of the other throughout its entire length. The best way to correct this condition is to cut two new openings at the bottom of each flue for two separate cleanout doors. Use the center cleanout door to fill in the partition with concrete block and mortar. The corrected chimney is shown in Fig. 2-9.

HOUSE REPAIRS

While on the subject of chimney repairs, this is a good time to discuss other repairs around the home that will directly affect your heating system. Cracks and crevices inevitably appear in every home, regardless of construction, and should be corrected. Besides allowing cold outside air into the home and heated air out, moisture, insects, dust, and dirt also enter. The sealing of these cracks and crevices should be on your list of repairs.

Caulking

Caulking compound, the "magic" material that clings to wood, stone, or metal, is probably all you will need to repair most cracks unless they are over ½ inch wide. Then use *oakum* or mortar first

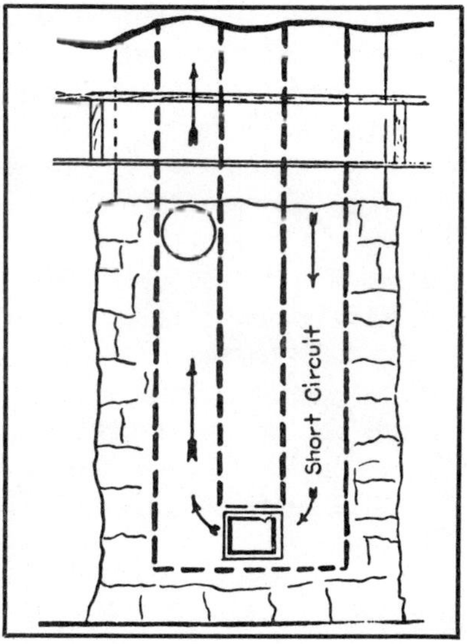

Fig. 2-8. A common situation is the use of one cleanout door for two separate flues. The problem should be corrected for best results and use of the chimney.

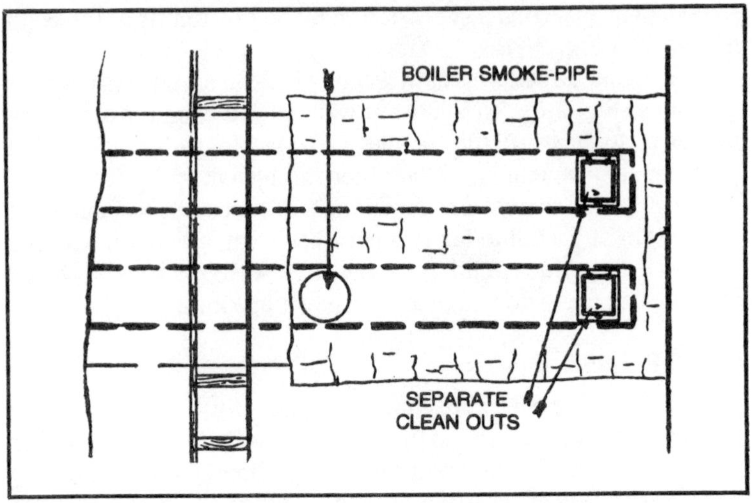

Fig. 2-9. Separate cleanout doors and the partition between filled in with mortar will correct the situation as indicated in Fig. 2-8.

before caulking. Cartridges of caulking compound with built-in nozzles can be slipped quickly into a top loading caulking gun. Disposal of empty cartridges is very easy.

Generally you will need to caulk around the chimney between the foundation and siding, around frames of windows and doors, around pipes where they penetrate walls (like an outside water faucet), and especially where a wooden structure joins masonry walls. A putty knife or similar tool should be used to pry away old compound before applying the new caulking.

Caulking should be an annual job even if your home is relatively new. Your home may appear to be free from cracks which might tempt you to go a few years without checking too closely. When cracks are obvious, you're going to have a tough job. An annual caulking will mean very minor work each year, and you will keep your wood or coal supply to the bare minimum.

Many homeowners choose to caulk their homes in the fall just prior to the cold season. I prefer to look for cracks and correct them in the early spring before too much foliage appears on vines, bushes, and the like. You have a better chance of finding all cracks and crevices. In the fall, even though frost has killed the foliage, there will still be dead leaves and other growths to hinder your inspection and work.

While on the subject of sealing cracks, you should also check your window panes every year. If you find that some of the putty needs replacing, remove the old putty with an old chisel or *glazier's knife*. Provide an airtight seal around the pane with a thin bead of glazing compound or putty (Fig. 2-10). String the putty between your thumb and forefinger, and wipe it into the rabbets in which the pane fits. Then use the tip of a putty knife to smooth it.

Weather Stripping

Weather stripping is equally as important as caulking for saving fuel and insuring your family's comfort in winter. The cracks around windows and doors of an average house, if not weather stripped, leak as much air as a 3 foot-square hole in your exterior wall.

Weather stripping comes in several different types-felt, wood and felt, metal, and metal and felt. All of these will do the job, but the all metal types are usually the easiest to apply. Simply reel the stripping out of its container, cut it to fit with scissors, and nail it in place (the nails are usually included).

Metal strips can be installed on windows by opening the window completely and inserting the weather stripping in the track where the window sash runs. Install the strip so it barely clears the window stop when the window is shut, compressing the springiness of the metal strip. Nail stripping to the sill to seal the gap under the sash when the window is closed.

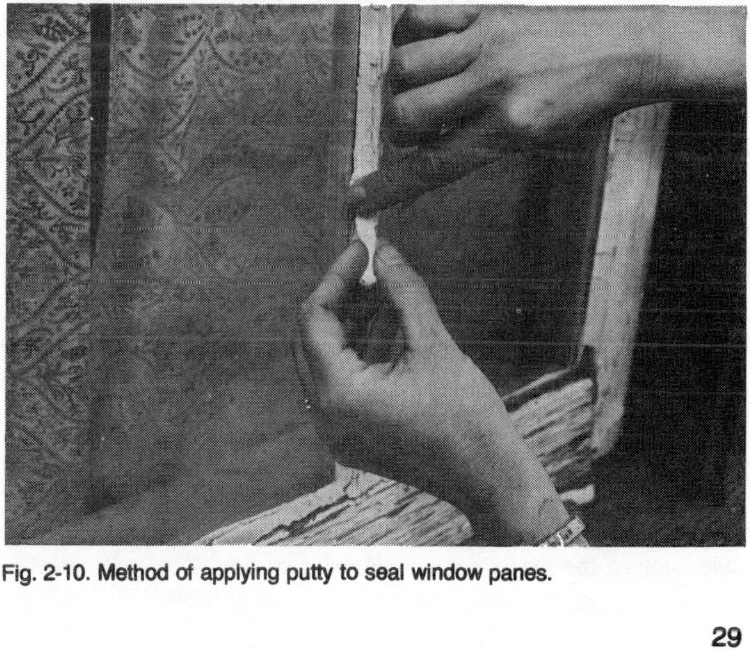

Fig. 2-10. Method of applying putty to seal window panes.

All exterior doors should be weather stripped in similar fashion. The door bottom is often sealed with special metal and felt weather stripping instead of ordinary types. The direction of opening and the clearance between the door and threshold determines whether this piece is screwed to the inside or outside of the door.

Insulation

Insulation is certainly not new. Man has used natural substances for centuries to keep heat in or out of his living quarters. The rough thatched huts of northern Europe and the British Isles had thick straw-covered roofs, while huts in Africa and the South Seas were (and are) covered with hollow sea grass and reeds for insulation.

Building experts agree that insulation is vital to holding down heating and cooling bills. Its importance to realty values is recognized by all housing authorities.

To save heat loss in winter and cut down on heat gain in summer, insulate the attic, floor, and walls with the proper amount of quality insulation. Attic spaces in both new and existing homes should be insulated with a minimum of 4 inches of insulation. Six inches is better. Some homeowners even use batts of fiber glass insulation 10 inches thick, especially when the home is electrically heated. If the attic space is open and readily accessible, fiber glass batts or loose insulation poured from a bag may be used. Many homes have flooring in the attic which prevents using batts or poured insulation. Blown insulation is the answer. Removing a few floorboards and using an insulation blower will enable you to fill the space between the ceiling and the attic floor with insulation.

Most insulation dealers will loan or rent you a small portable blower when you buy insulation. They'll also provide you with complete instructions on its use. Be sure to wear a respirator when installing insulation.

New homes are normally insulated by insulation board on the outside of the studs (between the studs and the outside finish) and 4-inch-thick fiber glass batts between the studs. In existing homes with wood siding, one of the top siding boards is removed around the perimeter of the house. Loose insulation is either poured or blown into the space between the studs. Alternatively, small holes can be drilled through the siding at the top of each between-stud cavity to accommodate the nozzle of the blowing machine. Then the siding is carefully replaced, or wooden plugs are used to plug the holes before the surface is refinished.

If the house has masonry outside the studs, there are still several ways in which the sidewalls can be insulated; the method depends on the structural characteristics of the house. Some homeowners drill down through the partition plates in their attics and then blow insulation down into the partitions through these holes. Since most homes have bridging or fire stops installed about halfway down between studs, the blown insulation will not completely fill the partition. In order to fill this space, the baseboard along the outside walls can be removed. Holes are drilled in the walls behind the baseboard, and insulation is blown through these holes. You may be able to blow the insulation up from the basement. Since the loose insulation is somewhat compressed as it is being blown in, very little will fall back down. All holes should be plugged and sealed after the job is finished.

Fiber glass batts are best for insulating under floors. Purchase the type with paper backing and of the correct width; the batts can be secured to the floor joists with a staple gun. Figures 2-11, through 2-13 show various types of house construction and how they should be insulated.

The cost of insulating your home will very quickly be returned in fuel savings. From then on, it's money in the bank (your account,

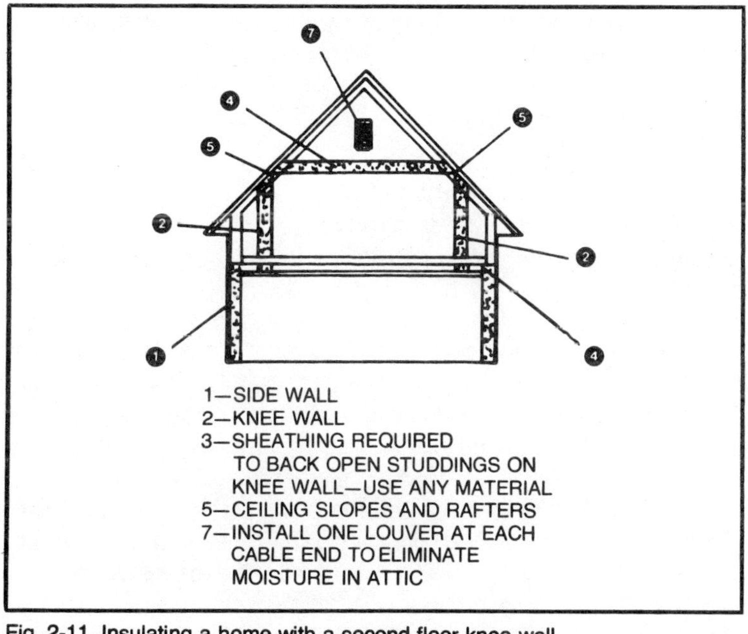

Fig. 2-11. Insulating a home with a second floor knee wall.

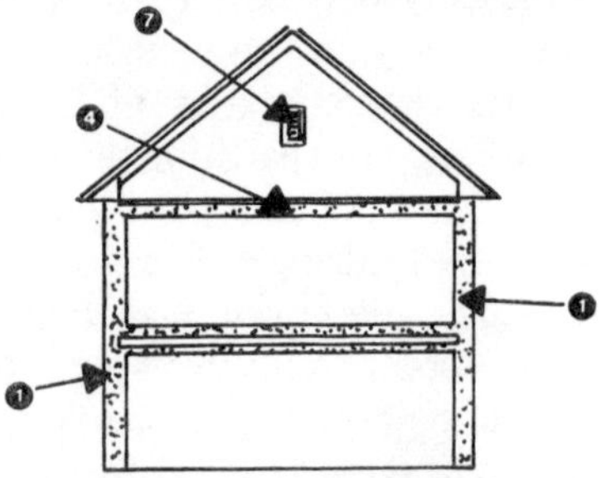

Fig. 2-12. Insulating a home with a vented, unheated attic.

not the fuel company's). Living comfort plus the increased resale value of your home after insulating are additional returns on your investment.

Many studies have shown that most of a home's heat loss occurs through an uninsulated roof. Recent studies using infrared cameras suggest that loss through doors and windows may be equally as significant.

Storm Windows and Doors

Sealing cracks around frames of windows and doors will help reduce heat loss. The addition of storm windows and doors or double glass will do wonders in cutting your fuel bills. Some homes' fuel bills have been cut in half by a complete storm window and door treatment.

Double glazing—two sheets of glass separated by a dead air space—is most easily incorporated in a new home, but you may order double glazed frames to fit in your present window openings. It's far less expensive to add conventional storm windows. If your windows are a nonstandard size, storm windows can be quickly and inexpensively ($10-18 per window) made to fit.

Should you want to do something immediately about the heat loss through your windows, you can purchase inexpensive snap-in clear plastic storm windows that are installed on the inside of your windows. No special tools are needed as the self-adhesive backing holds the mounting trim to the window frame. Most windows come in kits which include a plastic sheet, vinyl mounting trim for silled windows, and easy to follow instructions. If the standard sizes will not fit your windows, the trim may be cut with a regular handsaw. The plastic shield is cut with a pointed utility tool.

To add conventional storm/screen insulating windows, measure the window opening from the outside. Width is the distance between side casings. Height is the distance from the window sill to the top of the casing. Allow 1/16 inch or a little more clearance to compensate for out of square frames and to ensure easy installation.

When installing, make sure you don't distort the shape of the frame. The sides must remain straight and parallel. Some windows come with spring clips which hold the frame rigidly against the storm insert; the clips should not be removed until installation is complete. If your windows don't have such clips, check the fit of the insert by moving it up and down in the frame before you secure the frame's second side.

Most windows are predrilled to accept small wood screws. The screws will attach to the window casing more easily if you drill starter holes, using the holes in the storm frame as a guide. If the frame holes are badly positioned for the window, drill new frame holes. The soft aluminum frame is easily drilled with a standard twist bit.

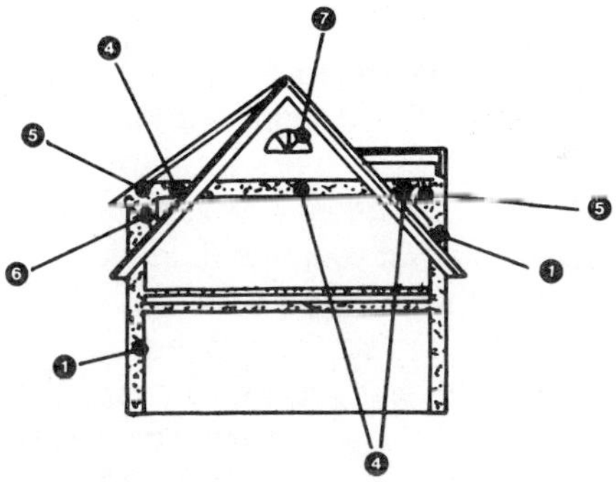

1—SIDE WALL
4—FLAT ATTIC—OPEN JOISTS
5—CEILING SLOPES ON RAFTERS
6—OUTSIDE WALL OF DORMER
7—LOUVER AT EACH END OF
ATTIC FOR VENTILATION

Fig. 2-13. Method of insulating a home with dormers.

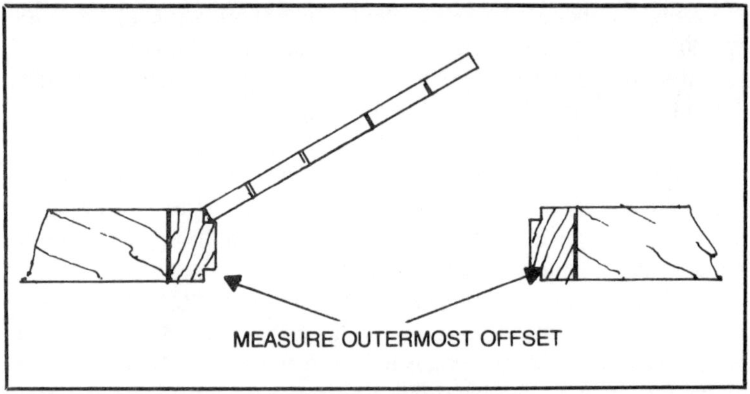

Fig. 2-14. Measuring a door opening to determine the size of the storm door.

Storm doors are installed similarly to the windows, but the door size should be measured as follows. Face the door opening from the outside. Measure the distance between the facing boards at the outermost offset (Fig. 2-14). Take measurements at the extreme top and bottom. Then measure the height of the door from the bottom of the top facing board or brick molding to the top of the threshold; this measurement should be made at both sides. Use the smaller of the two widths and height measurements when ordering the new door.

Siding Repair

The sealing of minor structural cracks can be accomplished using caulking compound. Wood siding works loose over the years, too; this siding should be repaired to keep the heat loss and gain to a minimum.

Renail all loose siding on your home. Replace any siding that is rotten or damaged. If there are still slight gaps, use caulking compound or some other type of sealer on them. This sealing will not only reduce your fuel bills, but you will also find that your home is easier to keep clean since less dust and dirt will enter.

Vapor Barriers

Now that your home is insulated and sealed up tight, your heating and cooling bill should drop considerably. There is one problem that you are going to encounter if you don't take precautions—excessive moisture. Efficient insulation can create a moisture problem. All the cracks and leaks that allowed heat to escape in an uninsulated home also carried away the excess moisture. With these cracks filled, the moisture remains inside your home.

There are several solutions to the problem of excessive moisture, the best being the use of vapor barriers which limit the moisture flow through your walls. In new construction, vapor barriers (usually sheets of *polyethylene*) should be installed as follows.

- All walls and floors exposed to outside temperatures should have a separately applied vapor barrier.
- The vapor barrier should be located on the warm or heated side of the insulation; this applies to all surfaces.
- The vapor barrier should be secured to a firm support at all openings such as windows, doors, electrical outlets, etc.
- A vapor barrier should be installed over dirt floors of crawl spaces and basements.
- A vapor barrier should be installed under all on grade and below grade slab floors.
- All joints should overlap at least 3 inches.

In a home where it is impossible to install a separate vapor barrier, paints that retard the flow of moisture can be used on the inside room surfaces. Reputable dealers can show you figures illustrating the relative ability of various paints to resist moisture penetration.

Exhaust Fans

There are also mechanical devices that will help provide your home with moisture control. *Exhaust fans* like the ones used in your bathroom or kitchen can be used to control moisture. The fans should be vented to the outside of the structure through the roof or wall and be equipped with an automatic damper.

The fans should be controlled by an automatic *humidistat* connected in parallel with the manual fan switch. When so connected, you may turn the fan on or off at any time when the humidistat is not calling for the removal of moisture. If the humidistat is set for 35-40 percent relative humidity and moisture content becomes greater than this amount, the fan will operate automatically, overriding the manual control.

During summer months, the humidistat should be set at 100 percent relative humidity or at the off position in order to prevent the fan from running continuously. You may also purchase a dehumidifier for removing excessive moisture, but if the recommendations given in the preceding paragraphs are followed, this should not be necessary.

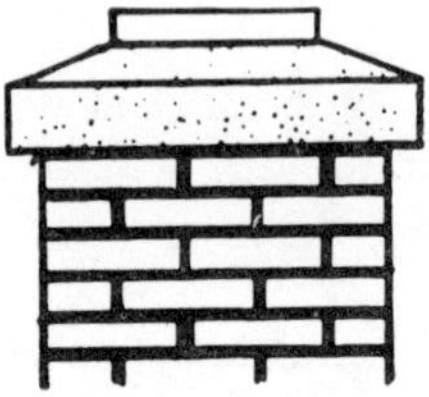

Chapter 3
Chimney Cleaning

The best method of keeping the chimney clean is to burn only seasoned wood in your wood stove or fireplace. This may not always be possible or even desirable. This chapter shows you some methods of cleaning chimneys used by professional chimney sweeps.

The main reason for a clean chimney is to help prevent chimney fires caused by creosote buildup within the chimney. Burning only seasoned wood will help cut down on the amount of creosote that is formed, but some creosote will always be formed regardless of the type of wood or the warmth of the chimney.

CHEMICAL CLEANING

I can still remember my grandmother, while cooking at her woodburning kitchen range periodically throwing a handful of *table salt* into the firebox. This action helped to keep the creosote buildup down. Actually, the table salt (*sodium chloride*) gave off fumes as it hit the hot fire. These fumes rising through the chimney caused the creosote to flake and fall from the chimney walls into the ashpit in the bottom of the chimney. By this method, the chimney was kept reasonably free from creosote buildup. During the 20 years I visited my grandparents at their farm, there was never a chimney fire. As far as I know, their chimneys were never cleaned by conventional methods.

There are several commercial products on the market today that are advertised to chemically clean chimneys. Most of these products consist mainly of *calcium chloride* or a mixture of *sodium chloride* and *copper sulfate.* You'll pay close to $10 a can for these chemicals. I doubt if they are much better than common table salt, but you may wish to give them a try.

Even if you use these chemicals (and burn seasoned wood), it is still advisable to give your chimney a good cleaning once a year. If you live in the country and have some evergreen trees close by, you can use a small cedar tree or a pine bough to obtain fairly good results.

CLEANING CHIMNEYS WITH EVERGREEN TREES

Obtain a small cedar tree or pine tree (practically any evergreen will do) that will fit snugly into your chimney. Secure a heavy weight onto the bottom of the trunk then tie a rope near the top of the tree long enough to lower the tree all the way down into the chimney. Be sure that the place where the rope is tied is strong enough to hold the weight of the tree as it is pulled back up through the chimney. The weights can be old window weights, small boat anchors, or similar items.

Once you have rigged up your tree, lower the butt end first down into the chimney, say 5 or 6 feet or so, and observe the tree as you move it up and down. If the tree fits too tightly, trim the tree now before you lower it further into the chimney. If the tree is too loose (doesn't scrub the chimney walls sufficiently), you'll have to cut a larger one.

When all is in order, lower the tree all the way down into the chimney (the weight tied to the bottom will aid in this operation). Now pull the tree up and let the weight pull it back down a few feet; pull it back up again. What you're trying to do is form a scrubbing motion with the tree rather than pulling it directly up. When you have reached the top of the chimney with this up and down scrubbing motion, lower the tree all the way down into the chimney again and repeat the operation. Continue to do so until the chimney is clean and free of creosote. You can obtain very good results by examining the chimney with a flashlight from the top, and a mirror and flashlight from the bottom.

When performing this cleaning operation, be sure to seal off the bottom of the chimney to keep soot and dirt out of your house. This also holds true for any mechanical cleaning operation.

MECHANICAL CHIMNEY CLEANING

A rope may be used to lower and raise tire chains inside a chimney to dislodge creosote and soot. A similar mechanical method is the use of a burlap sack filled with straw or chicken wire to make it fit snugly against the chimney walls. Either of these objects may be pulled up and down through the chimney to dislodge creosote and soot.

The disadvantage of using these methods is that they will sometimes knock out mortar joints or damage flue lining—making the chimney leak smoke which may cause fire outside the chimney. After cleaning a chimney using these means, carefully examine the chimney's interior with a flashlight. If any of the mortar joints have been damaged, they must be repaired.

If the mortar joints between the flue joints are in need of repair, fill a canvas bag with sand so that it will fit tightly inside the flue, yet loose enough to slide up and down. Lower the bag by a rope or chain down to a point just below the joint that needs repair. Pour a small amount of mortar down the chimney which will rest on the top of the canvas bag; have it well distributed over the bag so that it will touch the sidewalls. Then work the bag up and down to squeeze it into the joint. Should several joints in the chimney flue need repair, start with the lowest one first. Then work your way upward to the highest joint. This way you won't damage already repaired joints during the process.

The mortar used for chimney repairs should be a mixture of *portland cement,* sand, and water. It is used where the work is exposed to the action of the weather and when freezing weather is a problem. Another popular type of mortar is cement-lime mortar which is a mixture of portland cement, lime, sand, and water. It is generally recommended for masonry carrying heavy loads, for fireplaces and flues, and for locations where the work is partly protected. Cement-lime mortar costs less than cement mortar but more than the lime mortar. Besides being cheaper than the cement mortar, its greater plasticity makes it easier to be squeezed between flue joints.

CLEANING WITH WIRE BRUSHES

Specially designed wire brushes are usually recommended for cleaning chimneys. They can scrub out every particle of soot and creosote that can possibly be removed from a flue—usually without any damage to the flue lining or its joints. Opinions vary on the type of brush to use: square, round and hard, soft and long, short and weighted, unweighted, etc. Any of these brushes will do a good job of cleaning a chimney. Your main concern is to be thorough.

The use of wire brushes is the traditional approach used by

professional chimney sweeps. To begin the cleaning procedure, first close off all openings from the chimney that enter into the living area of the house—fireplace openings, stovepipe openings, and the like. Use a sheet of heavy paper taped completely over the opening. Be sure that not even a trickle of soot can leak into the house.

Using a safe ladder, climb up on the roof carrying a rope and weighted chimney brush (Fig. 3-1). The heavy brush is then inserted into the chimney, lowered all the way to the bottom, and then pulled back up to the chimney's top. Using this up and down action, continue scrubbing the inside of the flue until all particles of soot, ash, creosote, and other substances have been knocked loose.

When the chimney is clean, leave the roof, come back inside the house, and finish the job from the bottom. Allow enough time for the dust to settle. Then remove the tape and paper covering from the openings that you taped previously. Clean out the smoke chamber and smoke shelf (in the case of fireplaces) and/or the bottom chamber of the chimney by shoveling all the soot, ashes, and other debris into plastic trash bags. Dispose of the whole mess. If this is your first experience, the whole process will probably take you about two hours to complete. Once you get used to the procedure, however, you'll probably be able to complete the job in just over an hour.

Wire brushes are available at hardware stores and chimney sweeping supply centers. While they are expensive, purchase one if you use solid fuels for heating. A professional chimney sweep will usually charge $40 or more per flue.

Fig. 3-1. Timothy Johnson of Bentonville, VA, ties his 40-foot extension ladder for added safety as he carries equipment back and forth from the ground to the roof top.

Fig. 3-2. No, that's not Robin Hood on the roof with arrows. The shafts are fiber glass sections that screw together to form a cleaning rod to which steel brushes are screwed.

VACUUM CLEANING OF CHIMNEYS

Some professional chimney sweeps use a high volume vacuum cleaner which moves 700 cubic feet of air per minute, about 10 times the capacity of the average home vacuum cleaner. August West Systems, Box 663M, Westport, CT 06880 has done much to advance the art of sweeping chimneys. For around $1500, this company will send you a package of the specialized tools, operating instructions, and promotional ideas you'll need to start making money sweeping chimneys immediately. A black top hat also comes with the package for good luck.

One of the key elements of the August West System is the *soot sweeper* which is a highly specialized, high volume vacuum cleaner. Another unique part of this system is a set of specially developed fiber glass cleaning rods which snap together with *quick disconnect* similar to the connectors on socket wrenches or air-powered tools (Fig. 3-2). The dimensions of these rods are held to a size which makes them flexible enough to bend around smoke shelves, yet rigid enough to poke a wire chimney cleaning brush up a chimney. The construction of the rods makes them tough and highly resistant to fractures. In the unlikely event of a break, however, the snap-together rods splinter apart rather than snap off clean (leaving a brush stuck in a flue somewhere).

When using the August West System to clean the flue for a fireplace, spread a painter's drop cloth in front of the hearth to catch any debris that may accidentally escape the suction of the power vacuum cleaner. The inlet hose from the vacuum cleaner is then positioned in the back of the fireplace, and the machine is turned on. With the machine running, reach into the throat area of the fireplace and remove the damper. The vacuum cleaner will suck up most of the soot that falls during this operation.

Snap a brush onto one of the fiber glass cleaning rods, and push the brush up into the flue. Remember that the vacuum cleaner is still running. More rods are snapped onto the first one and fed up into the chimney until the brush pops out the top of the stack (Fig. 3-3). The

Fig. 3-3. The steel brush is screwed onto one fiber glass section and pushed into the chimney, another piece is screwed onto the first section, and the wire brush is pushed further down the chimney. This procedure is repeated until enough sections have been joined to reach the bottom of the chimney.

chimney is then scrubbed from the top down as described previously. The only difference is the chimney sweep remains conveniently and safely indoors at the base of the chimney rather than high on the roof, and the cleaning motion is from top to bottom rather than in the opposite direction.

As the brush is pushed up and pulled down, a shower of ashes, soot, and other debris falls into the fireplace, but this should present no problem. The vacuum sweeper will remove the material (Fig. 3-4). This brushing is continued until the debris is knocked loose from the flue lining and sucked into the vacuum cleaner.

Once the flue has been thoroughly cleaned, use a small brush to clean up the smoke chamber and smoke shelf. All of the fine dust is removed by the vacuum sweeper. Big clumps that are too heavy to be drawn into the hose should be shoveled or swept into paper shopping bags. You should be able to complete the entire process within 30 minutes.

Some companies are promoting vacuum systems that supposedly clean an entire chimney alone—making cleaning a flue with a brush obsolete. According to good authorities, the vacuum cleaner hasn't yet been made that can completely substitute for the hard work which a chimney sweep puts into a job.

Another place to obtain chimney sweeping supplies is Black Magic Chimney Sweeps, Box 977M, Stowe, VT 05672. This company sells manuals and equipment along with franchises and a training course. You may also want to purchase *The Homeowners' Resource* Book on the art of Sweeping Chimneys. It's available for about $2 from The Chimney Sweep Guild, c/o Krista Associates, Box 1175, Portland, ME 04104. You will also find helpful chimney sweeping hints in the Mother Earth News available at your local news stand or from P.O. Box 70, Hendersonville, NC 28739.

CALLING IN A PROFESSIONAL CHIMNEY SWEEP

If you decide to clean your own chimney, you will be working at 20 and 30-foot heights or more (Figs. 3-5 and 3-6). If you're uneasy when at these heights, you might call a professional to do the job for you. Look in the Yellow Pages of your telephone directory under "chimney cleaning." The better chimney cleaners will probably be members of the *Chimney Sweep Guild* and carry plenty of liability insurance. A typical ad is shown in Fig. 3-7.

When you have the sweep on the phone, describe the job and try to get a firm price. Call around and try to get the best deal. You usually get what you pay for, though, so the lowest bidder may not be your best bet. You'll also normally do better on price if you have the job done in spring or summer rather than fall or winter. Most people have their chimneys cleaned in fall or winter. When a sweep has more work than he can do, you're going to pay top dollar when he finally gets around to your home.

Fig. 3-4. Once all the sections have been joined, the fiber glass cleaning rod—with brush attached—is worked up and down the chimney flue until the soot and creosote loosen and fall to the bottom of the chimney. The chimney sweep then uses a vacuum cleaner at the bottom of the chimney to remove the loose debris from the chimney.

Fig. 3-5. Cleaning chimneys often means climbing at great heights.

While you're negotiating with the sweep, try to find out his level of knowledge. Ask specifically what work is included in the price and exactly how he goes about the job. He will probably tell you that he guarantees no mess in your living area, but be sure to ask him if he doesn't. You'll also want to know about his training for the job. There are correspondence courses that train the art of chimney cleaning. Did the sweep train in school, take a correspondence course, or study as an apprentice under another sweep? Experience is the main thing, so ask how many years he has been in business and approximately how many chimneys he has cleaned prior to yours.

The kind of equipment the sweep uses is another way to tell what kind of a job you'll be getting. Black Magic and August West are both good systems.

Try to give all the assistance you can during the sweep's appointment. Have all the furniture out of the way. Laying newspaper down around the area will also help the sweep do a cleaner job in less time. The going rate these days is between $30 and $50 per flue, but it's worth it if you get a thorough job.

OTHER CHIMNEY CLEANING METHODS

In 1928 when most rural residents used woodburning stoves exclusively for heat, the Department of Agriculture issued a bulletin describing how to clean chimneys without the aid of a chimney sweep. When the fires were started in the fall and the chimney failed to draw properly because of an accumulation of soot, the following remedies were recommended to cure the problem. You may or may not choose to use them for your own chimney cleaning.

Using Salt

Common salt is one of the simplest materials to use (as discussed earlier in this chapter). Before being applied, the salt should be dried and the fire put into good condition with a substantial bed of hot fuel on top. Sprinkle about a pound of salt on the fire close to the firebox or furnace door before shutting the door. Open all dampers so the fire will burn as rapidly as the drafts will permit. Keep the

Fig. 3-6. Keeping your balance while carrying cleaning equipment to the top of the chimney requires confidence and skill.

Fig. 3-7. Look in the Yellow Pages of your telephone directory under "chimney cleaning" for ads such as the one shown.

dampers wide open until the fumes have disappeared, or approximately 30 minutes. The soot is acted upon by the sodium and chlorine, which are dissociated when the salt is heated, and disappears as a gas or falls from the flue lining as scales. Additional salt may be needed to remove heavy deposits of soot. An occasional application of salt throughout the heating season will keep the flues clean.

Using Zinc

Another material sometimes used instead of salt is metallic *zinc;* a sheet about 6 to 8 inches square is generally effective. The jacket on a dry cell battery is composed of zinc, and for this reason beneficial results are secured by the common practice of burning old batteries on a hot bed of coal. This practice has been discouraged in modern times, since some batteries will explode when heated. If you have some safe method of removing the interior contents of zinc-lined dry cell batteries and burning only the lining, this is one inexpensive way to reduce the formation of creosote in your flue.

Chapter 4
Building Your Own Chimney

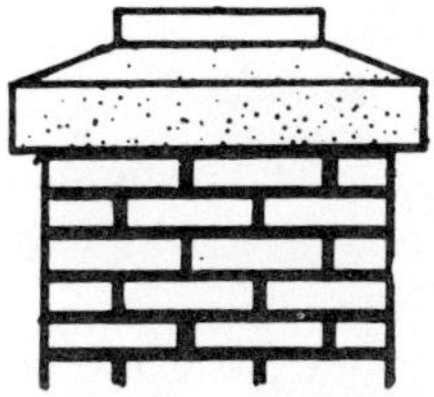

All fireplaces and fuel burning equipment like stoves and furnaces require some type of chimney (Fig. 4-1). The chimney must be designed and built so that it produces sufficient draft to supply an adequate quantity of fresh air to the fire, and expels smoke and gases emitted by the fire or equipment.

A chimney located entirely inside a building has better draft than an exterior chimney. The masonry retains heat longer when protected from cold outside air.

FLUE SIZE

The *flue* is the passage in the chimney through which the air, gases, and smoke travel. Proper construction of the flue is important. Its size (area), height, shape, tightness, and smoothness determine the effectiveness of the chimney in producing adequate draft and in expelling smoke and gases. Soundness of the flue walls may determine the safety of the building should a fire occur in the chimney. Overheated or defective flues are one of the chief causes of house fires.

Manufacturers of fuel burning equipment usually specify chimney requirements, including flue dimensions, for their equipment. Follow their recommendations.

HEIGHT

A chimney should extend at least 3 feet above flat roofs and at least 2 feet above a roof ridge or raised part of a roof within 10 feet of the chimney (Fig. 4-2). A hood (Fig. 4-2C) should be provided if a chimney cannot be built high enough above a ridge to prevent trouble from eddies caused by wind being deflected from the roof. The open ends of the hood should be parallel to the ridge.

Low cost metal pipe extensions are sometimes used to increase flue height, but they are not as durable or attractive as *terra-cotta* chimney pots or extensions. Metal extensions must be securely anchored against the wind and have the same cross sectional area as the flue. They are available with a metal cowl or top that turns with the wind to prevent the air from blowing down the flue.

SUPPORT

The chimney is usually the heaviest part of a building. It must rest on a solid foundation to prevent differential settlement in the building.

Concrete *footings* are recommended. They must be designed to distribute the load over an area wide enough to avoid exceeding the safe load bearing capacity of the soil. They should extend at least 6 inches beyond the chimney on all sides. The footings should be 8 inches thick for one-story houses and 12 inches thick for two-story houses having basements. If there is no basement, pour the footings for an exterior chimney on solid ground below the frost line.

If the house wall is of solid masonry at least 12 inches thick, the chimney can be built integrally with the wall. Instead of being carried down to the ground, it can be offset from the wall enough to

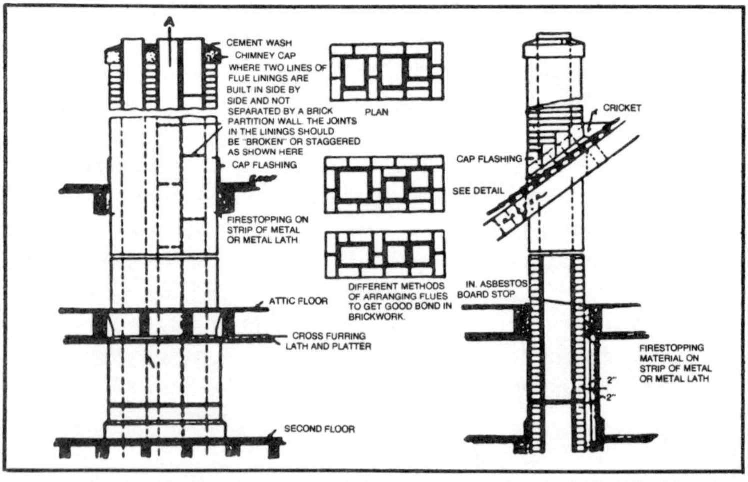

Fig. 4-1. Sectional view of a typical chimney — a requirement of all fuel burning equipment.

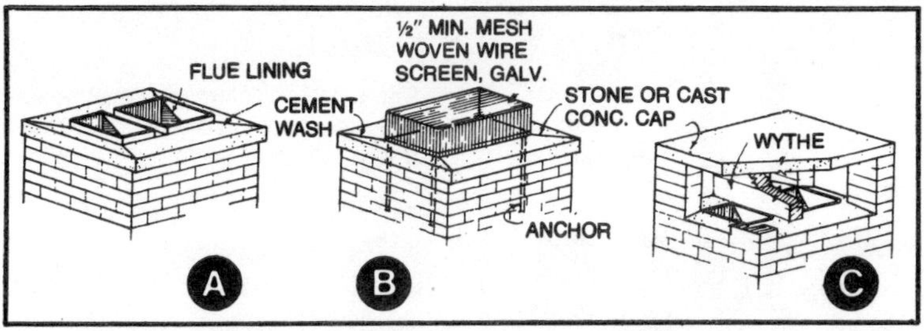

Fig. 4-2. Samples of chimney tops. (A) Acceptable way of finishing the top of a chimney; flue linings extend 4 inches above the cap. (B) Screen acts as spark arrester and also to keep birds and animals out during nesting season. C) Hood protects mortar around flue, keeps out rain, and helps prevent downdrafts.

provide flue space by *corbeling*. The offset should not extend more than 6 inches from the face of the wall, each course projecting not more than 1 inch, and should be not less than 12 inches high.

Chimneys in frame buildings should be built from the ground up, they can rest on the building foundation or basement walls if the walls are of solid masonry 12 inches thick and have adequate footings.

FLUE LINING

Chimneys are sometimes built without *flue lining* to reduce cost, but those with lined flues are safer and more efficient. Lined flues are definitely recommended for brick chimneys. When the flue is not lined, mortar and bricks directly exposed to the action of flue gases disintegrate. This disintegration plus that caused by temperature changes can open cracks in the masonry, which will reduce the draft and increase the fire hazard.

Flue lining must withstand rapid fluctuations in temperature and the action of flue gases. Therefore, it should be made of vitrified fire clay at least ⅝ inch thick.

Both rectangular and round-shaped linings are available. Rectangular lining is better adapted to brick construction, but round lining is more efficient.

Each length of lining should be placed in position—set in cement mortar with the joint struck smooth on the inside. Then the brick is laid around it. If the lining is slipped down after several courses of brick have been laid, the joints cannot be filled. Leakage will occur. In masonry chimneys with walls less than 8 inches thick, there should be space between the lining and the chimney walls. This space should not be filled with mortar. Use only enough mortar to make good joints and to hold the lining in position.

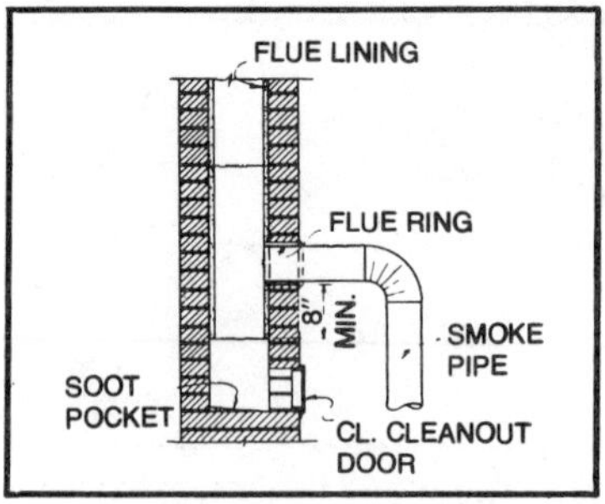

Fig. 4-3. Good chimney construction includes a soot pocket and cleanout door.

Unless it rests on solid masonry at the bottom of the flue, the lower section of lining must be supported on at least three sides by brick courses projecting to the inside surface of the lining. This lining should extend to a point at least 8 inches under the smoke pipe thimble or flue ring (Fig. 4-3).

Flues should be as nearly vertical as possible. If a change in direction is necessary, the angle should never exceed 45° (Fig. 4-4). An angle of 30° or less is better, because sharp turns set up eddies which affect the motion of smoke and gases. Where a flue does change directions, the lining joints should be made tight by mitering or cutting equally the ends of the adjoining sections. Cut the lining before it is built into the chimney; if cut after, it may break and fall out of place. To cut the lining, stuff a sack of damp sand into it. Then tap a sharp chisel with a light hammer along the desired line of cut.

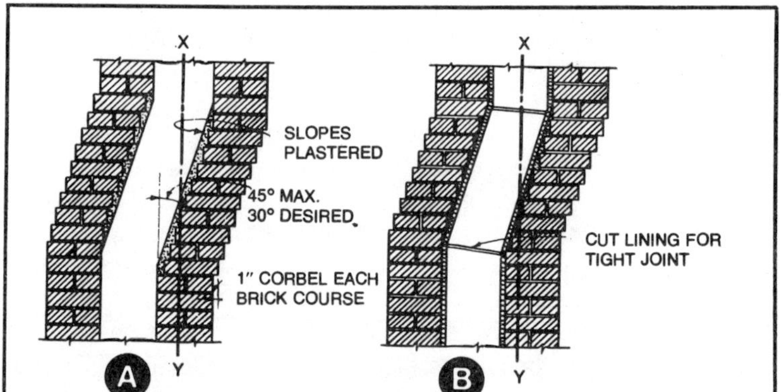

Fig. 4-4. For structural safety, the amount of offset must be limited so that the center line (X-Y) of the upper flue will not fall beyond the center of the wall of the lower flue. Start the offset of the left wall of an unlined flue two brick courses higher than the right wall, so that the area of the sloping section will not be reduced after plastering (A). The method of cutting lining to make a tight joint is shown in (B).

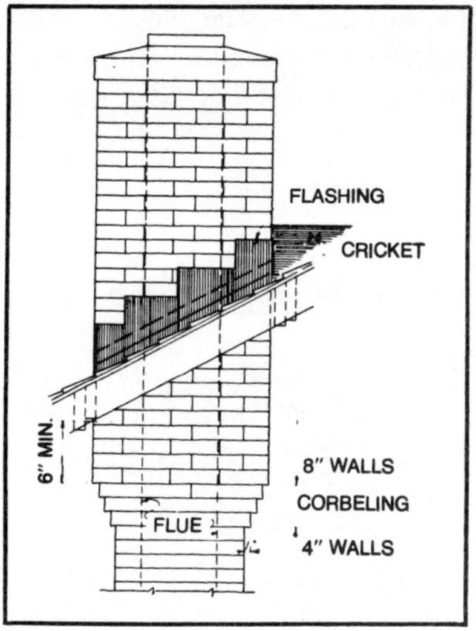

Fig. 4-5. Corbeling chimneys is one way to provide a thicker chimney where it extends through the roof to lessen the chance of openings developing, which can allow sparks to start fires in the woodwork or roofing.

When laying lining and brick, draw a tight fitting bag of straw up the flue as the work progresses to catch material that might fall and block the flue.

WALLS

Walls of chimneys with lined flues and not more than 30 feet high should be at least 4 inches thick if made of brick or reinforced concrete and at least 12 inches thick if made of stone. Flue lining is recommended especially for brick chimneys. The lining can be omitted, though, if the chimney walls are made of reinforced concrete at least 6 inches thick or of unreinforced concrete or brick at least 8 inches thick. A minimum thickness of 8 inches is recommended for the outside wall of a chimney exposed to the weather.

Brick chimneys that extend up through the roof may sway enough in heavy winds to open up mortar joints at the roof line. Openings to the flue at that point are dangerous, because sparks from the flue may start fires in the woodwork or roofing. Make the upper walls 8 inches thick by starting to offset the bricks at least 6 inches below the underside of roof joists or rafters (Fig. 4-5).

Chimneys may contain more than one flue. Building codes generally require a separate flue for each fireplace, furnace, or boiler. If a chimney contains three or more lined flues, each group of two flues must be separated from the other single flue or group of two flues by brick divisions or wythes at least 3-¾ inches thick (Fig. 4-6). Two flues grouped together without a dividing wall should have the lining joints staggered at least 7 inches. The joints must be completely filled with mortar. If a chimney contains two or more unlined flues, the flues must be separated by a well bonded wythe at least 8 inches thick.

MORTAR

Brickwork around chimney flues and fireplaces should be laid with cement mortar. It is more resistant to the action of heat and flue gases than lime mortar.

A good mortar to use in setting flue linings and all chimney masonry, except firebrick, consists of 1 part portland cement, 1 part hydrated lime (or slaked lime putty), and 6 parts clean sand, measured by volume. Firebrick should be laid with *fireclay*.

SOOT POCKET AND CLEANOUT

A *soot pocket* and *cleanout* are recommended for each flue (Fig. 4-3). Deep soot pockets permit the accumulation of an excessive amount of soot which may take fire. Therefore, the pocket should be only deep enough to permit installation of a cleanout door below the smoke pipe connection. Fill the lower part of the chimney, from the bottom of the soot pocket to the base of the chimney, with solid masonry.

The cleanout door should be made of cast iron. It should fit snugly and be kept tightly closed to keep air out.

A cleanout should serve only one flue. If two or more flues are connected to the same cleanout, air drawn from one to another will affect the draft in all.

SMOKE PIPE CONNECTION

No range, stove, fireplace, or other equipment should be connected to the flue for the central heating unit. Each unit should be connected to a separate flue, because if there are two or more connections to the same flue, fires may occur from sparks passing into one flue opening and out through another.

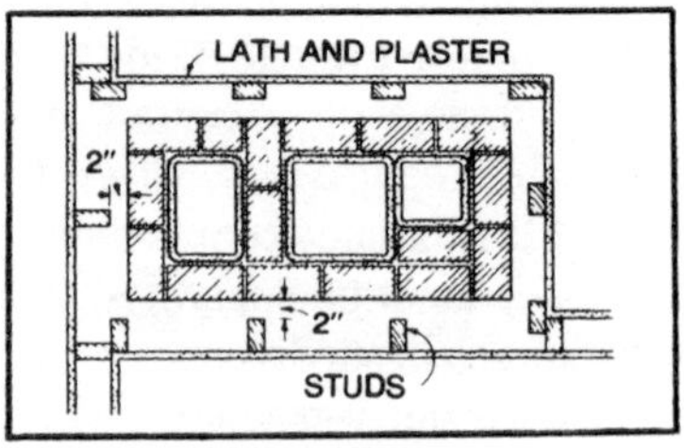

Fig. 4-6. Plan view of chimney showing proper arrangement of three flues. The division walls are bonded with sidewalls by staggering the joints of successive courses. All wood framing should be at least 2 inches from any brickwork of the chimney.

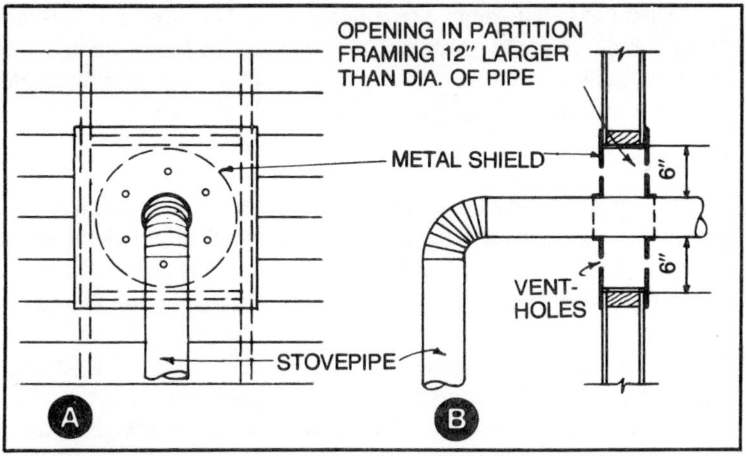

Fig. 4-7. When a smoke pipe passes through a wood partition, either cut an opening in the partition and insert a galvanized iron, double wall ventilating shield at least 12 inches larger than the pipe, or install at least 4 inches of brickwork of other incombustible material around the pipe.

Smoke pipes from furnaces, stoves, or other equipment must be correctly installed and connected to the chimney for safe operation. A smoke pipe should enter the chimney horizontally and should not extend into the flue (Fig. 4-3). The hole in the chimney wall should be lined with fireclay, or metal thimbles should be tightly built into the masonry. (Metal thimbles or flue rings are available in diameters of 6, 7, 8, 10, or 12 inches and in lengths of 4½, 6, 9, and 12 inches.) To make an airtight connection where the pipe enters the wall, install a closely fitting collar and apply boiler putty, good cement mortar, or stiff clay.

A smoke pipe should never be closer than 9 inches to woodwork or other combustible materials. If it is less than 18 inches from woodwork or other combustible materials, cover at least the half of the pipe nearest the woodwork with fire-resistant material. Commercial fireproof pipe covering is available.

If a smoke pipe must pass through a wood partition, the woodwork must be protected. Either cut an opening in the partition and insert a galvanized iron, double wall ventilating shield at least 12 inches larger than the pipe (Fig. 4-7) or install at least 4 inches of brickwork or other incombustible material around the pipe.

Smoke pipes should never pass through floors, closets, or concealed spaces. They should never enter the chimney in the attic. Each summer when they are not in use, smoke pipes should be taken down, cleaned, wrapped in paper, and stored in a dry place.

When not in use, smoke pipe holes should be closed with tight fitting metal flue stops. Don't use papered tin. If a pipe hole is to be abandoned, fill it with bricks laid in good mortar. Such stopping can be readily removed if necessary.

INSULATION

No wood should be in contact with the chimney. Leave a 2-inch space (Fig. 4-6) between the chimney walls and all wooden beams or joists (unless the walls are of solid masonry 8 inches thick, in which case the framing can be within ½ inch of the chimney masonry).

Fill the space between wall and floor framing with porous, nonmetallic, incombustible material such as loose cinders (Fig. 4-8). Do not use brickwork, mortar, or concrete. Place the filling before the floor is laid. It not only forms a fire stop but prevents the accumulation of shavings or other combustible material.

Flooring and subflooring can be laid within ¾ inch of the masonry. Wood studding, furring, or lath should be set back at least 2 inches from chimney walls. (Plaster can be applied directly to the masonry, but this is not recommended because settlement of the chimney may crack the plaster.) A coat of cement plaster should be applied to chimney walls that will be encased by a wood partition or other combustible construction.

If baseboards are fastened to plaster that is in direct contact with the chimney wall, install a layer of fireproof material, such as asbestos, at least ⅛ inch thick between the baseboard and the plaster (Fig. 4-8).

CONNECTION WITH ROOF

Where the chimney passes through the roof, provide a 2-inch clearance between the wood framing and the masonry for fire

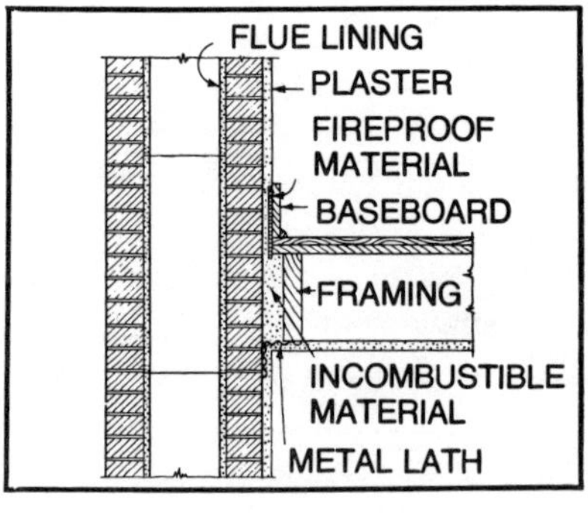

Fig. 4-8. Method of insulating wood floor joists and baseboard at a chimney.

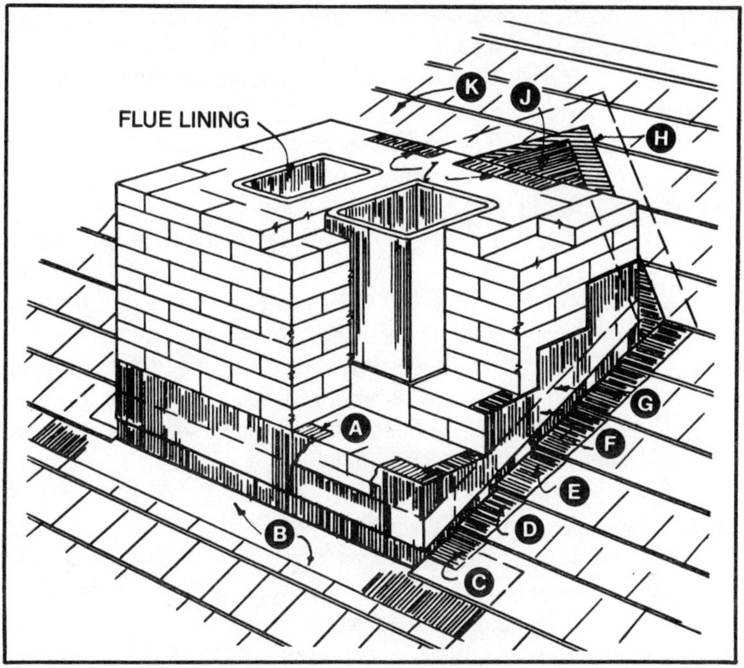

Fig. 4-9. Correct method of flashing around a chimney where it passes through a roof. Sheet metal (H) over cricket (J) extends at least 4 inches under the shingles (K) and is counterflashed at (L) in joint. Base flashings (B,C,D, and E) and cap flashings (A,F, and G) lap over the base flashings to provide watertight construction. A full bed of mortar should be provided where the cap flashing is inserted in joints.

protection. The clearance will permit expansion due to temperature changes, settlement, and slight movement during heavy winds.

Chimneys must be flashed and counterflashed to make the junction with the roof watertight (Figs. 4-9 and 4-10). When the chimney is located on the slope of a roof, a cricket (J in Fig. 4-9) is built high enough to shed water around the chimney (Fig. 4-11). Corrosion-resistant metal such as copper, zinc, or lead should be used for flashing. Galvanized or tinned sheet steel requires occasional painting.

TOP CONSTRUCTION

Figure 4-2A shows a good method of finishing the top of the chimney. The flue lining extends at least 4 inches above the cap or top course of brick and is surrounded by at least 2 inches of cement mortar. The mortar is finished with a straight or concave slope to direct air currents upward at the top of the flue and to drain water from the top of the chimney.

Hoods (Fig. 4-2C) are used to keep rain out of chimneys and to prevent downdraft due to nearby buildings, trees, or other objects. Common types are the arched brick hood and the flat stone or cast concrete cap. If the hood covers more than one flue, it should be divided by wythes so that each flue has a separate section. The area of the hood opening for each flue must be larger than the area of the flue.

Spark arresters (Fig. 4-2B) are recommended when burning fuels that emit sparks, such as sawdust, or when burning paper or other trash. They may be required when chimneys are on or near combustible roofs, woodland, lumber, or other combustible materials. They are not recommended when burning soft coal, because they may become plugged with soot.

Spark arresters do not entirely eliminate the discharge of sparks, but if properly built and installed they greatly reduce the hazard. They should be made of rust-resistant material and should have screen openings not larger than ⅝ inch or smaller than 5/16 inch. (Commercially made screens that generally last for several years are available.) They should completely enclose the flue discharge area and must be securely fastened to the top of the chimney. The arresters must be kept adjusted in position, and they should be replaced when the screen openings are worn larger than normal size.

SMOKE TEST

Every flue should be tested as follows before being used and preferably before the chimney has been furred, plastered, or other-

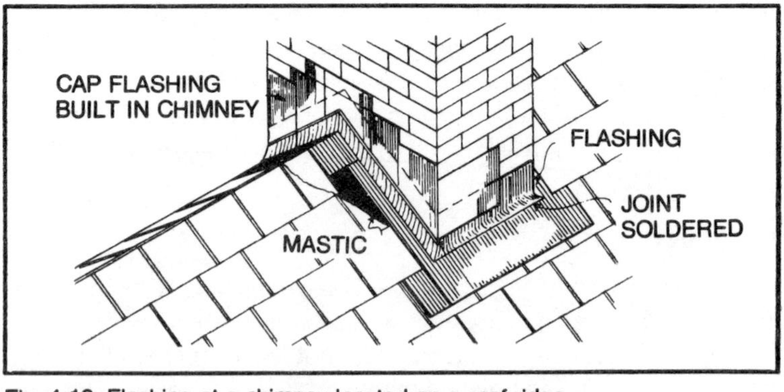

Fig. 4-10. Flashing at a chimney located on a roof ridge.

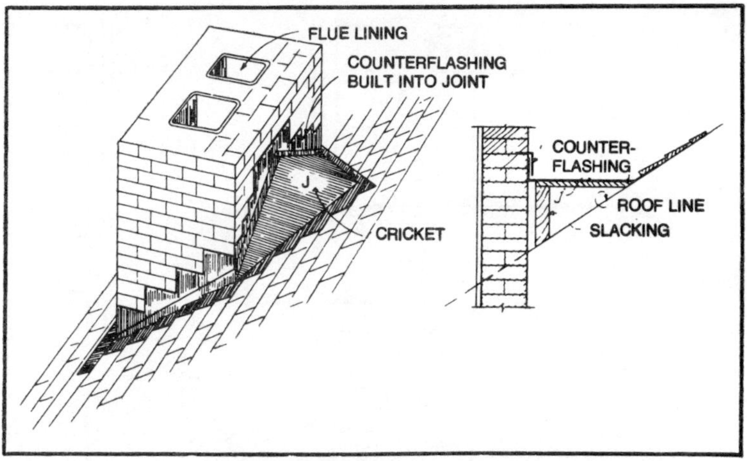

Fig. 4-11. Typical cricket showing how counterflashing is built into joints.

wise enclosed. Build a paper, straw, wood, or tar paper fire at the base of the flue. When the smoke rises in a dense column, tightly block the outlet at the top of the chimney with a wet blanket. Smoke that escapes through the masonry indicates the location of leaks.

This test may show bad leaks into adjoining flues, through the walls, or between the lining and the wall. Correct defects before the chimney is used. Since such defects may be hard to correct, you should check the initial construction carefully.

The foregoing description of fireplace construction is designed to aid you in determining whether you would like to attempt such a project or not. The information to follow will go into more detail which, if followed, should allow you to construct a safe workable chimney of your own.

HOW A CHIMNEY WORKS

You already know that the main purpose of a chimney is to create a draft which, in turn, carries smoke to the outside of a building. Draft in a chimney flue is caused by the difference in weight between a volume of the gases formed by the fire and an equal volume of air on the outside. Since air expands as it is heated, it becomes lighter in weight, causing the pressures at the base of the flue to become unequal. The lighter gases (heated air) rise in the chimney, tending to equalize the pressure. This pressure or suction causes air to be drawn through the fuel beds in the wood stove or fireplace which in turn helps to keep the fire burning. For this reason, chimneys must have certain specified heights. Further-

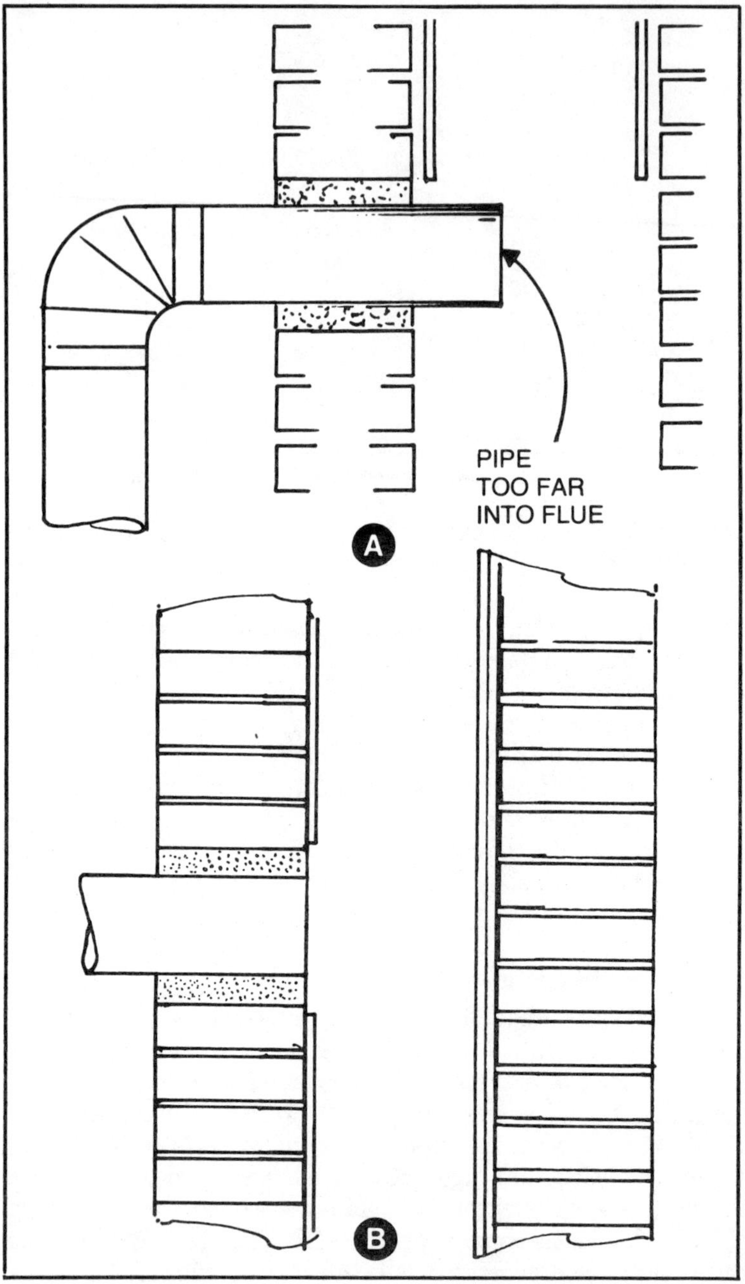

Fig. 4-12. The stovepipe in (A) protrudes too far into the chimney flue, while (B) shows the correct way of installing a stovepipe.

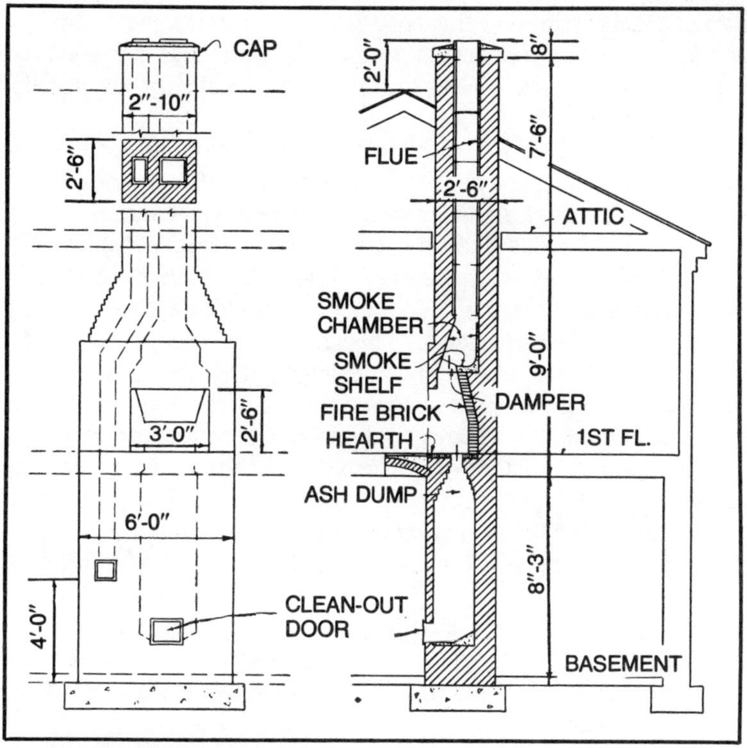

Fig. 4-13. Note the cleanout door.

more, the chimney walls must be airtight—having no cracks in the flue to allow any leakage of air. As mentioned previously, the flue should be as straight as practicable and have a few inside projections as possible.

Contrary to popular belief, a round flue is more efficient than a square or rectangular flue because the gases and smoke rise with a circular swirling motion. A round flue, therefore, offers less resistance than the other types. A square or oblong flue is not effective over its entire cross sectional area, because the column of smoke does not fill the corners as it rises. Then why are most residential chimneys either square or rectangular? Most chimneys did not contain flue lining prior to the twentieth century. They were constructed from various masonry materials and then plastered smooth on the inside with mortar cement to offer the least amount of resistance to smoke. Obviously, a rectangular or square-shaped chimney is easier to construct than a round one, so these designs were used for most residential chimneys. When flue liners came into use, this tradition remained. Square or rectangular chimneys are built for most residences today. The chimneys designed by consulting engineers, such as the ones used in industry, are usually round in shape.

Round flue liners are available and may be put to good use in the construction of new residential chimneys. The outside walls can remain square or rectangular in shape (to follow tradition), and a round or circular flue liner is used inside the chimney walls.

Chimneys may be constructed without flue liners if local building codes permit. Fireclay flue lining is superior for chimney construction because it is made of a special mixture of clays thoroughly burned to withstand fire and heat, changes in temperature, and all actions resulting from fire, smoke, gases, or acids. Another reason for using flue lining is to save on the cost of the masonry work. An unlined chimney that is only plastered on the inside generally requires chimney walls at least twice as thick as those chimneys that are lined. A flue lining also offers much less chance of sparks or flames going through the chimney walls than does a plain brick chimney.

A separate flue should be provided for each heating device in the home: fireplace, woodburning stove, furnace, etc. If more than one heating unit is connected to the same flue, the draft for each unit is reduced in half.

The size of flue required for a heating unit depends upon how much draft is needed and must be determined by the manufacturer of the heating equipment. The sectional area of a fireplace flue should be from one-tenth to one-twelfth the area of the fireplace opening. For example, a fireplace opening 42 inches wide and 30 inches high has an area of 1260 square inches. One-half of that is 105 square inches. A 12-inch circular flue lining has an actual inside area

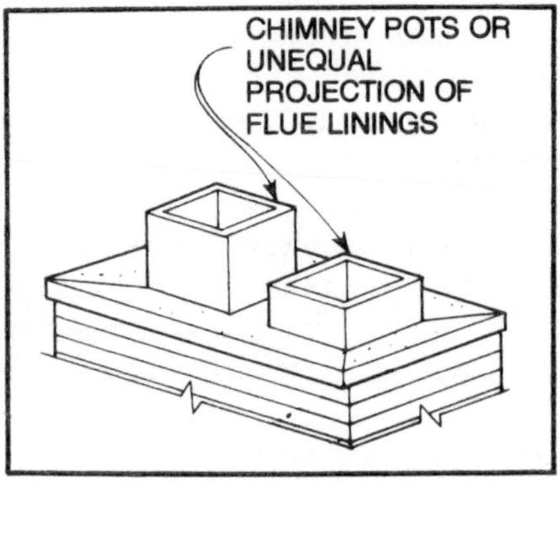

Fig. 4-14. Typical chimney cap.

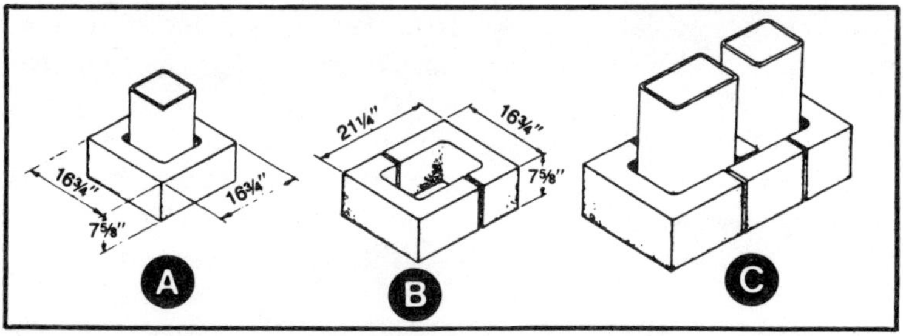

Fig. 4-15. Chimney blocks of various sizes.

of 113 square inches and would be satisfactory. A 13 × 13-inch flue lining, outside measurement, has a gross inside area of 127 square inches, but the effective area would be only 100 square inches.

The connections between smoke pipes and chimneys should be carefully built to assure proper draft and safety from fire. All openings for chimneys should consist of metal thimbles around which the chimney brickwork should be carefully laid and cemented. The pipe should not extend into the chimney (Fig. 4-12A). Such practice reduces the flue area and may result in a poor draft as would be caused by closing the damper in the smoke pipe. The correct installation is shown in Fig. 4-12B.

Although many building codes require basement ceilings to be plastered as a protection from fire, Fig. 4-13 illustrates the best practice for providing protection from excessive heat. Provision should be made for a soot pocket with a cleanout door at the bottom of each flue, other than for a fireplace. The top of the door should be placed as close as possible to, and not more than 1 foot from, the smoke entrance at the base of the chimney. It should fit snugly to prevent the admission of air (Fig. 4-13).

To prevent excessive downdraft in a chimney, the top should extend at least 3 feet above flat roofs and 2 feet above the ridge of a peak roof. The chimney should not be on the side of the house next to a tree or a building higher than itself, as these may cause eddies and force air down the chimney (Fig. 2-4).

Every chimney should be supported on a proper foundation, and no weight should be carried by a part of the building other than the walls when properly corbeled. The foundation and footings should be carefully proportioned to the weight of the chimney and the safe bearing capacity of the soil. In ordinary house construction it is sufficient to make the footing 6 inches wider than the chimney on all sides and 8 to 12 inches deep. In all cases the footing should start well below the frost line.

The tops of all chimneys must be capped with either stone or concrete to prevent weathering caused by rain and changing temperatures. Figure 4-14 shows a typical cap for a one flue chimney. Note that the flue lining extends above the top of the concrete cap from 2 to 4 inches. The flue lining is surrounded with cement mortar to a depth next to the flue lining of about 2 inches. This mortar is sloped from the sides of the flue to the edges of the concrete.

CONCRETE CHIMNEYS

Good chimneys, whether located on the interior or the exterior of the building, should always rest on solid concrete footings extended well below maximum frost depth. They should be of sufficient width and thickness to carry the weight of the chimney without danger of settlement. When the chimney forms a part of a building's wall, the chimney block must be securely bonded into it. Vitrified flue lining or firebrick flue lining, extending the entire height of the flue, is recommended by the National Board of Fire Underwriters.

The flue size should not be less than 8½ × 8½ inches for heating plants and 8½ × 12 inches for fireplaces. When two flues are

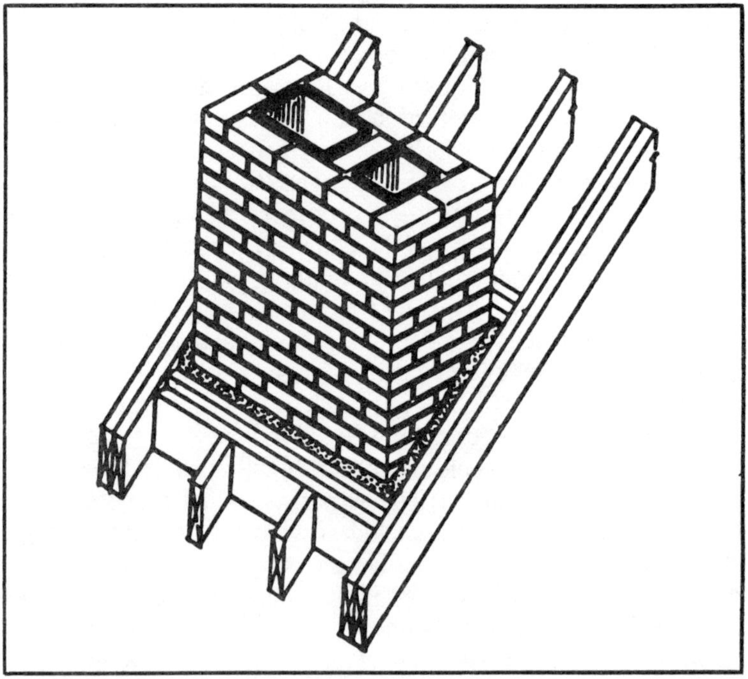

Fig. 4-16. Headers and trimmers used around the chimney.

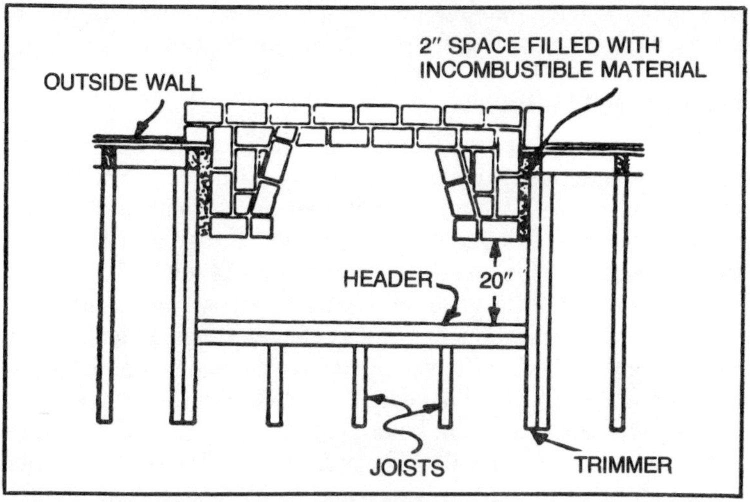

Fig. 4-17. Framing around the fireplace and chimney.

built in a chimney, the joints of the adjoining sections of flue linings should be staggered at least 7 inches. Where there are more than two flues in a chimney, at least every third flue should be separated by masonry at least 3-¾ inches thick and bonded into the masonry wall of the chimney. Masonry should be at least 8 inches thick on the exposed sides of chimneys built into exterior walls. Figure 4-15 illustrates chimney blocks of different sizes. Figure 4-15A shows a block with a 9-⅛ × 9-⅛-inch opening. The block has an opening of 9-⅛ × 13-⅝-inches.

FRAMING FOR CHIMNEYS

Wood framing has to be strengthened around chimneys when it is necessary to cut away part of the joist. This is done by doubling or tripling the joists on either side of the opening. These joists are called *header* and *trimmer* joists. Trimmers run parallel with the floor joists and headers at right angles (Fig. 4-16). The short joists which frame into the headers are known as *tail* joists. Where spans are short and loads are light, the framing around openings can be nailed together. If the headers are more than 6 feet in length, they should be fastened to their supporting joists by means of iron *stirrups* or joist hangers (Fig. 4-16).

Framing around chimneys (Fig. 4-17) is generally held 2 inches from the chimney. The space between is filled with incombustible material (Fig. 4-18). Wooden members must never frame into a chimney. A *pilaster* or corbel may be built as part of a chimney to support wooden framing members, however, provided there is plenty of brickwork around the flues and fireplaces.

FLASHING FOR CHIMNEYS

Wherever there are openings in a roof for a chimney, extra precaution must be taken to prevent leaks. The only satisfactory way of doing this is by using strips of sheet tin, galvanized iron, zinc, or copper. While flashing is usually metal, it may also be heavy impregnated felt (Fig. 4-19) or combinations of felt and metal.

Regardless of the material used, the details of flashing are practically the same. Copper is the preferred flashing metal for all roofs except for cedar shingles and metal roofs. The acid in cedar shingles attacks copper. For cedar shingles lead-coated copper, tin, or *terneplate* should be used. Where metal roofing is used, the same metal should be used for flashing as for the roofing. Although this discussion is based on the use of copper for flashing, remember that other metals are handled in about the same way.

Copper flashing is generally 16-ounce *roofing temper,* which means it weighs 16 ounces per square foot and is soft and pliable. Heavier copper, 18-ounce or 20-ounce, is advisable for better work. You can use 24-ounce copper for under flashings of mission tile.

Tin should be 1-C (about 30 U.S. gauge) or 1-X (about 28 U.S. gauge) thickness. Tin should only be used where exposed faces are accessible for the necessary frequent painting. Zinc may be used in place of copper for all flashing with the exception of cedar shingle roofs.

Lead or lead-coated copper may be used for any flashing, but its use generally is restricted to the most enduring roofs such as sheet

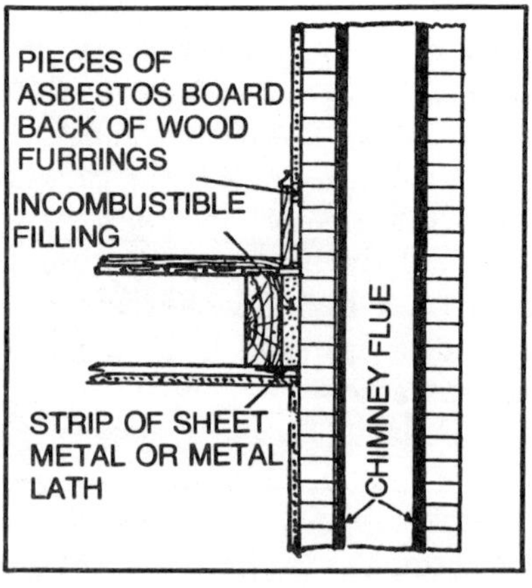

Fig. 4-18. Spaces between chimney and woodwork are filled with incombustible material.

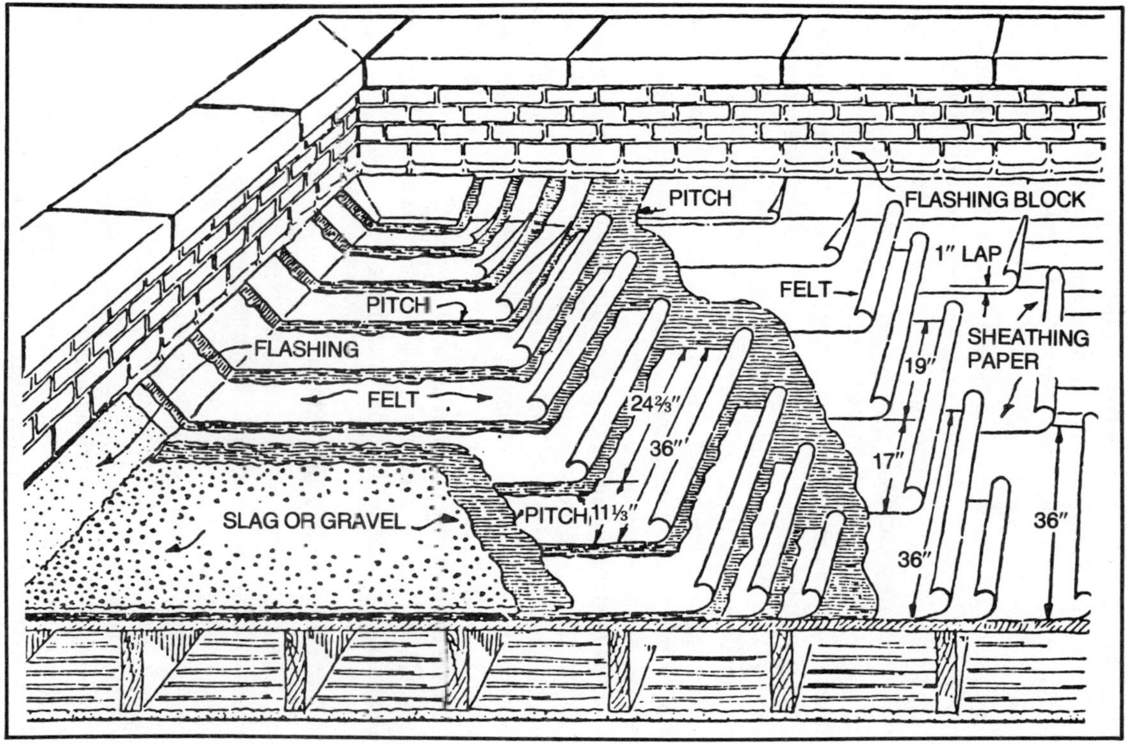

Fig. 4-19. Built-up roof using felt as flashing.

lead, slate, or clay tile. It is safer not to use lead where it is to be built into masonry walls, as lime or cement mortar will corrode the materials. When in doubt, you should consult the manufacturer or supplier. Aluminum generally is used for roof coverings of that material.

Where chimneys pass through a roof, the junction must be carefully flashed to prevent leaks. The flashing consists of two parts: the lower part or base flashing fitted into the shingles, and the upper or cap flashing built into the brickwork (Fig. 4-20). The lower part of the cap flashing folds down to lap 4 inches or more over the upper part of the base flashing. One piece of base flashing is required for each course of shingles. It is laid with its lower edge at just about the butt line of the succeeding course of shingles. The flashing should extend onto the roof at least 4 inches, and 5 or 6 inches is better. Cap flashing is usually 8 or more inches wide with its lower edge cut on a diagonal to fit the pitch of the roof. Cap flashing must be built into the chimney as the brickwork is laid and should extend into the mortar joint at least 3 inches. Many specifications require that it be built in the full 4 inches and be turned against the flue lining 1 or 2 inches. Note in Fig. 4-20 at the lower edge of the chimney that the base flashing becomes a single sheet or apron which rests on top of the shingles. The cap flashing is one continuous piece which laps down over the base flashing.

Included in flashing work is the *watershed* (sometimes called a *saddle* or *cricket*) on the high side of a roof back of a chimney or other similar penetration in the line of water flow (Fig. 4-20). Extend flashings under all unit roofings in valleys and at eaves to a point beyond which water, snow, or ice may collect. It is advisable to increase the size or extension of flashings wherever there may be reasonable doubt. Remember that a 4-inch extension is about the minimum in the most favorable situations.

ESTIMATING CHIMNEY MATERIALS

The usual method for determining the number of bricks required in a chimney is to find the number of bricks per foot of the chimney's height and multiply by the total height taken in feet. To find the number of bricks required per foot of height, you must first know how many bricks are required for one course of the chimney, and then how many courses there are in 1 foot in height.

Figure 4-21 shows various sizes of chimneys and flues and how the bricks are bonded in each case. For each size of flue and chimney, two layers of brick courses are shown. This is done to

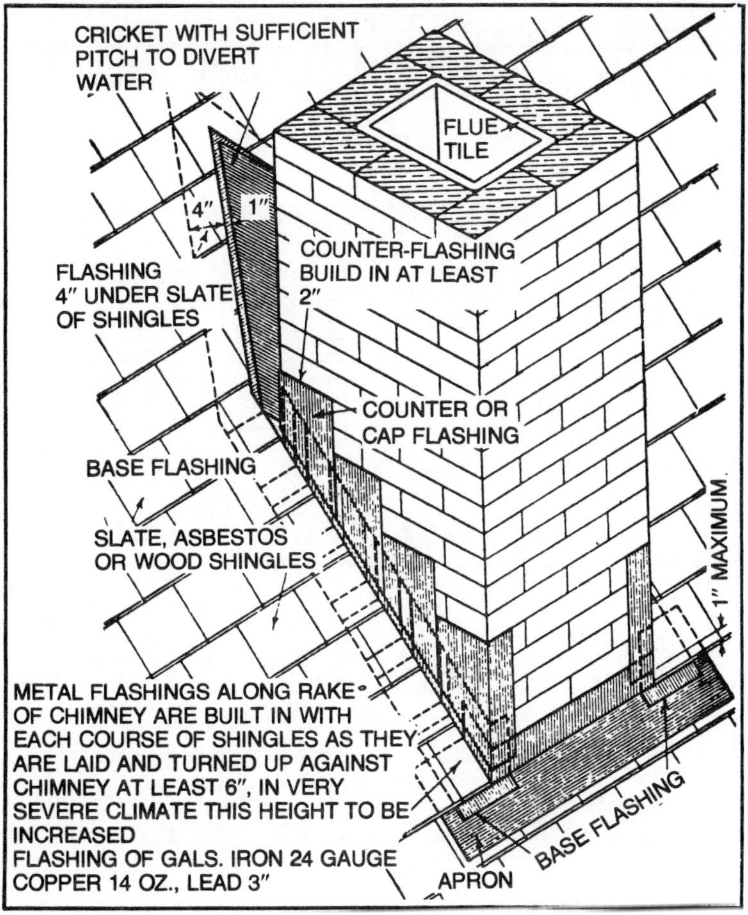

Fig. 4-20. Where chimneys pass through a roof, the junction must be carefully flashed to prevent leaks.

show how the courses will bond. In some cases one course will require more bricks than the other, as for example chimney "6." This will require 12 bricks for one course and 13 bricks for the other course. This makes an average of 12½ bricks per course.

You may also figure how many courses make up 1 foot in height. The standard brick is 2¼ inches thick and the average joint is ½ inch thick. This makes a 2-¾-inch height for every course. For 1 foot or 12 inches in height, there are $12 \div 2.75 = 4.36$ courses.

The chimney shown as 1 in Fig. 4-21 has 6 bricks to each course and requires 27 bricks per foot of height ($6 \times 4.36 = 26.16$). Chimney 6 in Fig. 4-21 has an average of 12½ bricks per course and

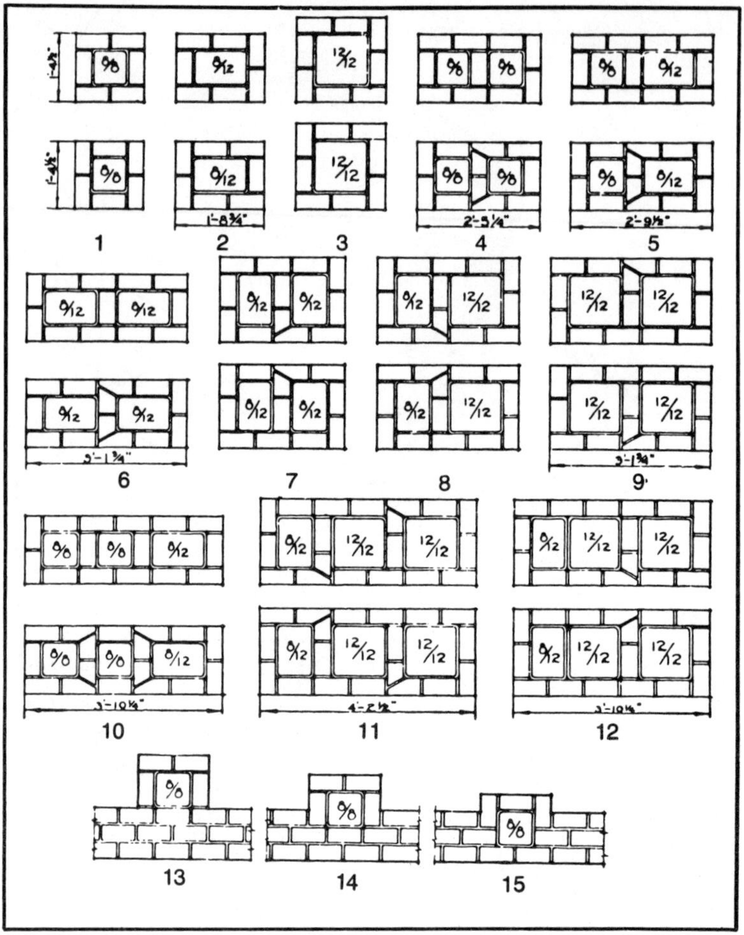

Fig. 4-21. Various sizes of chimneys and flues and how the bricks are bonded in each.

would require 55 bricks per foot in height (12½ × 4.36 = 54.5). To find the number of bricks in a chimney, multiply the bricks per foot of height by the total height. A chimney requiring 27 bricks per foot of height, if built 30 feet high, would require 810 bricks (30 × 27).

The chimneys shown by 13, 14, and 15 in Fig. 4-21 are built as part of a wall and do not require as many extra bricks. Type 13 in Fig. 4-21 requires four extra bricks for each course. Type 14 in Fig. 4-21 requires only two extra bricks—that is, two more bricks for each course than would be required if the wall were built straight without the chimney.

Chapter 5
Building Your Own Fireplace

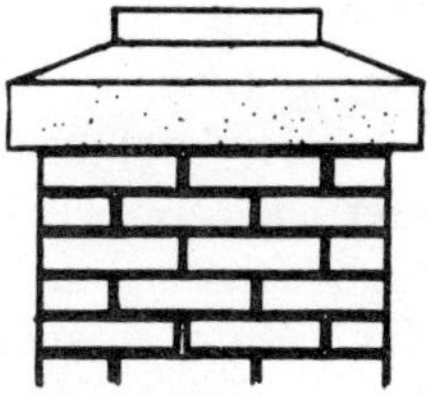

The worsening national energy problem has tempted many Americans to use their woodburning fireplaces again to help reduce the use of electricity, gas, or oil to heat their homes. Fireplaces are not as economical as one might imagine, and they are used more for their charm and aesthetics than for cutting fuel costs. Tests have indicated that fireplaces, as ordinarily constructed, are only about one-third as efficient as a good stove or circulator heater. Much of the energy content of the burning wood is lost up the chimney. A considerable volume of air heated with other sources is "pumped" out of the house and up the chimney by the largely uncontrollable fireplace draft. Glass fire screens with small draft holes at the bottom are helpful in reducing air loss up the chimney, but radiation from the fire is also reduced.

A well-designed, properly built fireplace can do the following:

- Provide additional heat.
- Provide all the heat necessary in mild climates.
- Enhance the appearance and comfort of the room.
- Burn certain combustible materials as fuel that otherwise might be wasted: scrap lumber, coke, and the like.

DESIGN

Varied fireplace designs are possible. Figures 5-1 through 5-3 show well-designed units. You'll find other designs in this chapter and many of the home and garden magazines.

A fireplace should harmonize in detail and proportion with the room in which it is located. Safety and utility, though, should not be sacrificed for appearance.

Location of the fireplace within a room will depend on the location of the existing chimney or the best location from the standpoint of safe construction for the proposed chimney. A fireplace should not be located near doors.

Fireplace openings are usually made from 2 to 6 feet wide, and the type of fuel available will suggest a practical width for yours. For example, where cordwood (4 feet long) is cut in half, an opening 30 inches wide is desirable. Where coal is burned, a narrower opening can be used.

Height of the opening can range from 24 inches for an opening 2 feet wide to a height of 40 inches for one that is 6 feet wide. The higher the opening, the more chance there is of a smoky fireplace. Dimensions of standard fireplace sizes are shown in Table 5-1.

Generally, the wider the opening is, the greater the depth. A shallow opening throws out relatively more heat than a deep one, but holds smaller pieces of wood. You have the choice, therefore, between a deeper opening that holds larger, longer burning logs and a shallower one that takes smaller pieces of wood but throws out more heat. In small fireplaces, a depth of 12 inches may permit good

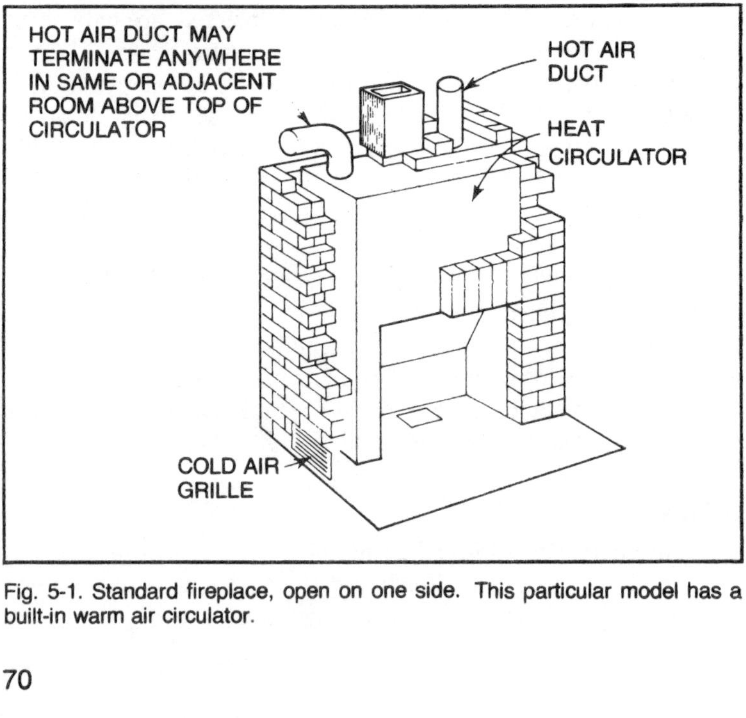

Fig. 5-1. Standard fireplace, open on one side. This particular model has a built-in warm air circulator.

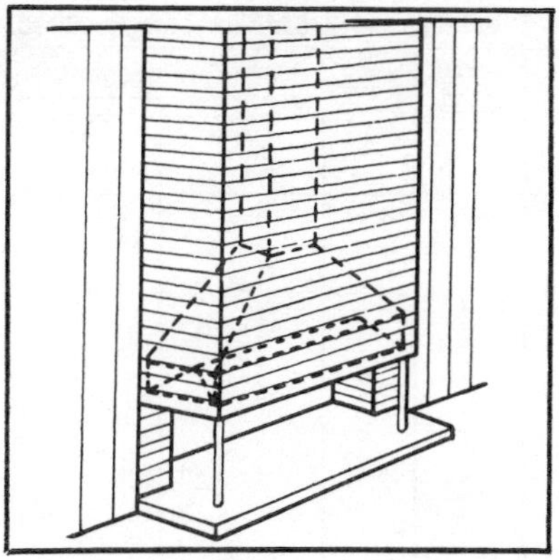

Fig. 5-2. Fireplace open on three sides.

draft, but a minimum depth of 16 inches is recommended to lessen the danger of brands falling out on the floor. Suitable screens should be placed in front of all fireplaces to minimize the danger from brands and sparks. Second-floor fireplaces are usually made smaller than first-floor ones because of the reduced flue height.

CONSTRUCTION

Fireplace construction is basically the same regardless of design. Figure 5-4 shows construction of a typical fireplace.

Footings

Foundation and footing construction for chimneys with fireplaces is similar to that for chimneys without fireplaces (see Chapter 4). Be sure the footings rest on good firm soil below the frost line.

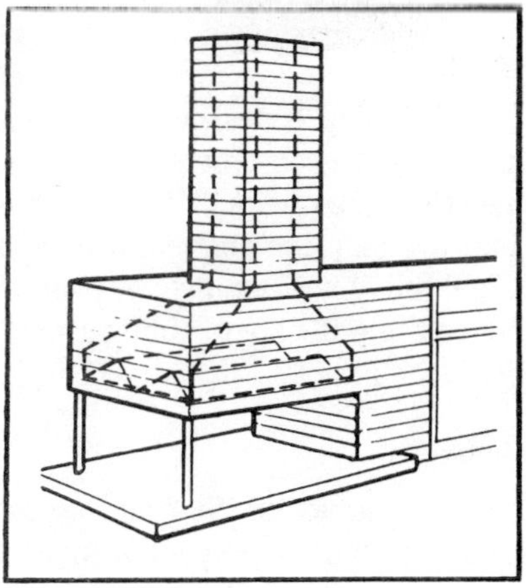

Fig. 5-3. Another fireplace designed with openings on three sides. This particular model is highly suited for use in the end of a room divider and similar situations.

Table 5-1. Recommended Dimensions for Fireplaces and Size of Flue Lining Required.

Size of Fireplace Opening		depth	minimum width of back wall	height of vertical back wall	height of inclined back wall	Size of Flue Lining Required	
width w	height h	d	c	a	b	standard rectangular (outside dimensions)	standard round (inside diameter)
inches	inches	inches	inches	inches	inches	inches	inches
24	24	16-18	14	14	16	8½ × 13	10
28	24	16-18	14	14	16	8½ × 13	10
30	28-30	16-18	16	14	18	8½ × 13	10
36	28-30	16-18	22	14	18	8½ × 13	12
42	28-32	16-18	28	14	18	13 × 13	12
48	32	18-20	32	14	24	13 × 13	15
54	36	18-20	36	14	28	13 × 18	15
60	36	18-20	44	14	28	13 × 18	15
54	40	20-22	36	17	29	13 × 18	15
60	40	20-22	42	17	30	18 × 18	18
66	40	20-22	44	17	30	18 × 18	18
72	40	22-28	51	17	30	18 × 18	18

Hearth

The fireplace *hearth* should be made of brick, stone, terra-cotta, or reinforced concrete at least 4 inches thick. It should project at least 20 inches from the chimney breast and should be 24 inches wider than the fireplace opening (12 inches on each side).

The hearth can be flush with the floor so that sweepings can be brushed into the fireplace, or it can be raised. Raising the hearth to various levels and extending in length as desired is presently a common practice, especially in contemporary design. If there is a basement, a convenient ash dump can be built under the back of the hearth (Fig. 5-5).

In buildings with wooden floors, the hearth in front of the fireplace should be supported by masonry trimmer arches or other fire-resistant construction. Wood centering under the arches used during construction of the hearth and its extension should be removed when construction is completed.

Walls

Building codes generally require that the back and sides of fireplaces be constructed of solid masonry or reinforced concrete at least 8 inches thick. They should be lined with firebricks or other approved noncombustible materials not less than 2 inches thick or steel lining not less than ¼ inch thick. Such lining may be omitted when the walls are of solid masonry or reinforced concrete at least 12 inches thick.

Jambs

The *jambs* of the fireplace should be wide enough to provide stability and to present a pleasing appearance. For a fireplace opening 3 feet wide or less, the jambs can be 12 inches wide if a wood mantel will be used, or 16 inches wide if they will be of exposed masonry. For wider fireplace openings, or if the fireplace is in a large room, the jambs should be proportionately wider.

Fireplace jambs are frequently faced with ornamental bricks or tile. No woodwork should be placed within 6 inches of the fireplace opening. Woodwork above and projecting more than 1½ inches from the fireplace opening should be placed not less than 12 inches from the top of the fireplace opening.

Lintel

A *lintel* must be installed across the top of the fireplace opening to support the masonry above. For fireplace openings 4 feet wide or

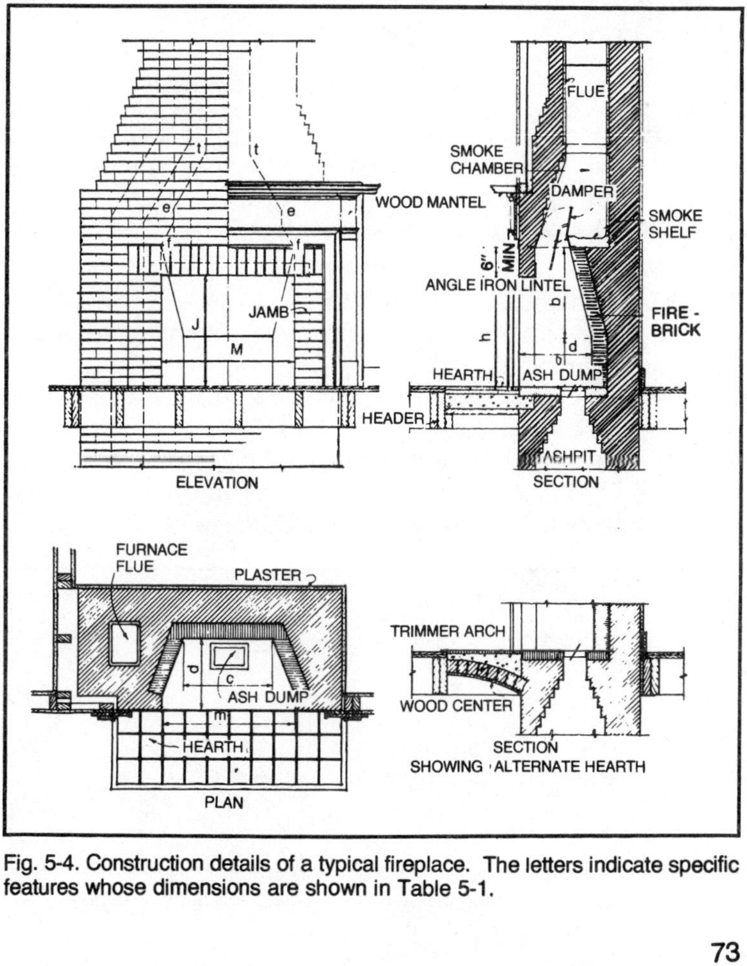

Fig. 5-4. Construction details of a typical fireplace. The letters indicate specific features whose dimensions are shown in Table 5-1.

less, ½ by 3-inch flat steel bars, 3½ by 3½ by ¼-inch angle irons, or specially designed damper frames may be used. Wider openings will require heavier lintels. If a masonry arch is used over the opening, the fireplace jambs must be heavy enough to resist the thrust of the arch.

Throat

Proper construction of the *throat* area is essential for a satisfactory fireplace. The sides of the fireplace must be vertical up to the throat, which should be 6 to 8 inches or more above the bottom of the lintel. The area of the throat must be not less than that of the flue; that is, the length must be equal to the width of the fireplace opening. The width will depend on the width of the damper frame if a damper is installed. Five inches above the throat, the sidewall should start sloping inward to meet the flue.

Damper

A *damper* consists of a cast-iron frame with a hinged lid that opens or closes to vary the throat opening. Dampers are not always installed, but they are definitely recommended, especially in cold climates. With a well-designed, properly installed damper, you can do the following:

- Regulate the draft.
- Close the flue to prevent loss of heat from the room when there is no fire in the fireplace.
- Adjust the throat opening according to the type of fire to reduce loss of heat. For example, a roaring pine fire may require a full throat opening, but a slow burning hardwood log fire may require an opening of only 1 or 2 inches. Closing the damper to a small opening will help reduce the loss of heat up the chimney.

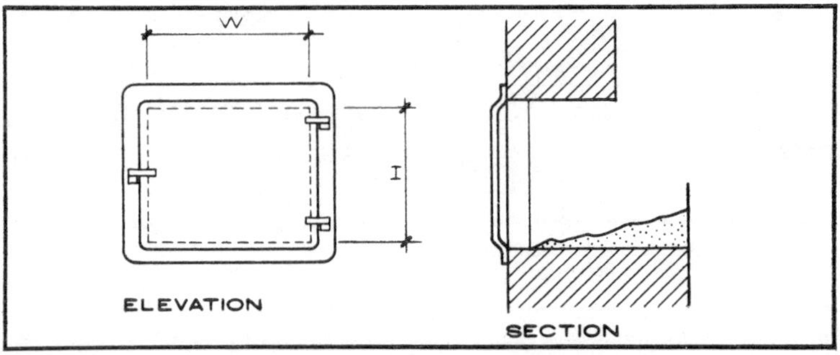

Fig. 5-5. Elevation and section drawings of a typical ash dump and door.

- Close or partially close the flue to prevent loss of heat from the main heating system. When air heated by a furnace goes up a chimney, an excessive amount of fuel may be wasted.
- Close the flue in the summer to prevent insects from entering the house through the chimney.

Dampers of various designs are on the market. Some are designed to support the masonry over fireplace openings, thus replacing ordinary lintels.

Responsible manufacturers of fireplace equipment usually offer assistance in selecting a suitable damper for a fireplace. It is important that the full damper opening equal the area of the flue.

Smoke Shelf and Chamber

A smoke shelf prevents down draft and is made by setting the brickwork at the top of the throat back to the line of the flue wall for the full length of the throat. Depth of the shelf may be 6 to 12 inches or more, depending on the depth of the fireplace.

The smoke chamber is the area from the top of the throat to the bottom of the flue. The sidewalls should slope inward to meet the flue. Both the smoke shelf and the smoke chamber walls should be plastered with cement mortar at least ½ inch thick.

Flue

Proper proportion between the size (area) of the fireplace opening, size (area) of the flue, and height of the flue is essential for satisfactory operation of the fireplace. For example, the area of a lined flue 22 feet high should be at least one-twelfth of the area of the fireplace opening. The area of an unlined flue or a flue less than 22 feet high should be one-tenth the area of the fireplace opening.

Table 5-1 lists dimensions of fireplace openings and, in the last two columns, indicates the size of flue lining required for a given size fireplace opening and also the size of opening to use with an existing flue. Flue construction principles given in Chapter 4 apply also to fireplace flues.

MODIFIED FIREPLACES

Modified fireplaces are manufactured fireplace units. They are made of heavy metal and designed to be set in place and concealed by the usual brickwork or other construction. They contain all the essential fireplace parts: the firebox, damper, throat, smoke shelf, and chamber. In the completed installation, only grilles or heat registers show. Figure 5-6 shows one of the several designs of modified fireplaces available. This type of fireplace offers two advantages over conventional ones. The correctly designed and proportioned firebox provides a ready-made form for the masonry, which reduces the chance of faulty construction and assures a smokeless fireplace. When properly installed, the better designed units heat more efficiently than ordinary fireplaces. They circulate heat into the cold corners of rooms and can deliver heated air through ducts to upper or adjoining rooms.

The use of a modified fireplace unit can increase the cost of a fireplace (although manufacturers claim that labor, materials, and fuel saved offset any additional costs). It should not be necessary to use one merely to insure an attractive, well-proportioned fireplace. You can build an equally attractive and satisfactory masonry fireplace by careful construction techniques.

Even a well-designed modified fireplace unit will not operate properly if the chimney is inadequate. Proper chimney construction is as important for these units as it is for ordinary fireplaces.

PREFABRICATED FIREPLACES AND CHIMNEYS

Prefabricated fireplace and chimney units—all parts needed for a complete fireplace to chimney installation—are on the market. One such unit is shown in Fig. 5-7. Such units offer these features:

- Wide selection of styles, shapes, and colors.
- Pretested design that is highly efficient in operations.
- Easy and versatile installation—can be installed freestanding or flush against a wall in practically any part of a house.
- Light in weight.
- Lower cost than comparable masonry units.

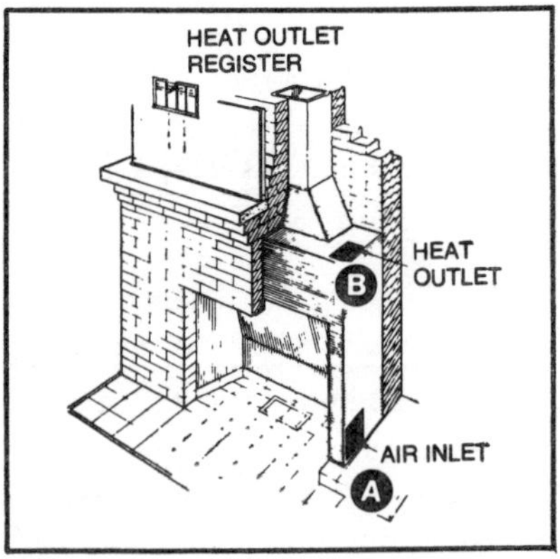

Fig. 5-6. A modified fireplace. Air is drawn through inlet (A) from the room being heated. It is heated by contact with the metal sides and back of the fireplace. Air rises by natural circulation. It is discharged through outlet (B). The inlets and outlets are connected to registers which may be located at the front (as shown) or ends of the fireplace, or on the wall of an adjacent or second-story room.

The basic part of the prefabricated fireplace is a specially insulated metal firebox shell. Since it is light in weight, it can be set directly on the floor without the heavy footing required for masonry fireplaces.

Prefabricated chimneys can be used for furnaces, heaters, and incinerators as well as for prefabricated fireplaces. The chimneys are tested and approved by Underwriters Laboratories and other nationally recognized testing laboratories. They are rapidly being accepted for use by building codes in many communities.

OUTDOOR FIREPLACES

Outdoor fireplaces range from simple makeshift units to elaborate structures designed to harmonize with the enhance and appearance of the house and the landscape. Plans for outdoor fireplaces are available from various sources. Magazines frequently feature articles on outdoor fireplace design and construction. If the fireplace is to be built with local labor and material, a relatively simple design is advisable.

Built-in features such as ovens, cranes, grilles, storage compartments, sinks, and benches add to the appearance and convenience of fireplaces. Dealers in outdoor fireplace equipment usually have catalogs listing types and sizes of accessory equipment. Elaborately designed fireplaces that include many built-in features or are an integral part of a building should be built by skilled labor.

Low level fireplaces are made of brick, stone, cement blocks, slate, and similar masonry materials with the grill placed on or near the top (Fig. 5-8). This type of cooking unit can be located almost anywhere and can even be incorporated into a brick or stone wall. This is the type of fireplace often seen in picnic areas set up by the National Park Service.

About the only disadvantage of the low level fireplaces is the necessity of having to bend over when turning a steak or placing and removing any other type of food. The extra visibility that it allows can be worthwhile in some backyard settings. Due to its low profile, this type of fireplace will not distract viewers from perhaps more pleasant scenery around your yard.

Other types of outdoor fireplaces include combination oven/grill units, smoke ovens, and large fireplaces with counter-height grills. Great diversity in design can be obtained with the larger types of fireplaces. For example, grills can vary in size. They can be either stationary or movable. Counter space for serving and preparing food can be built right into the unit—on either or both sides.

Ovens can be built under the counter space, or this area can be used for storage of barbecue gloves, charcoal, charcoal starter, etc.

If the grill is lotted or messed, no chimney is required (although recommended). All solid metal cooking plates must be provided with a chimney.

Elaborately designed fireplaces that include many built-in features such as sinks and benches are becoming common. Many are an integral part of a building. Employ skilled labor when attempting the larger, more complicated designs.

Planning

Although many outdoor fireplace designs seem simple and can be built in little time by most handymen, proper planning is required prior to beginning the construction. First, check with your local building inspector to find out the regulations pertaining to building outdoor fireplaces. You should also check with your local fire marshal.

Always have a drawing of the unit you plan to build—either sketched or drawn to scale. A materials list should then be made in order to determine an estimated cost. The location of the fireplace should be your next consideration, and this can greatly affect the size and type of fireplace that can be built.

The location of the fireplace is governed by several factors: fire regulations, building codes, practicality, and aesthetics. From a practical point of view, the closer a fireplace is to the house, the easier it will be to transport food and cooking utensils to and from the outdoor cooking area. Care should be taken not to locate the

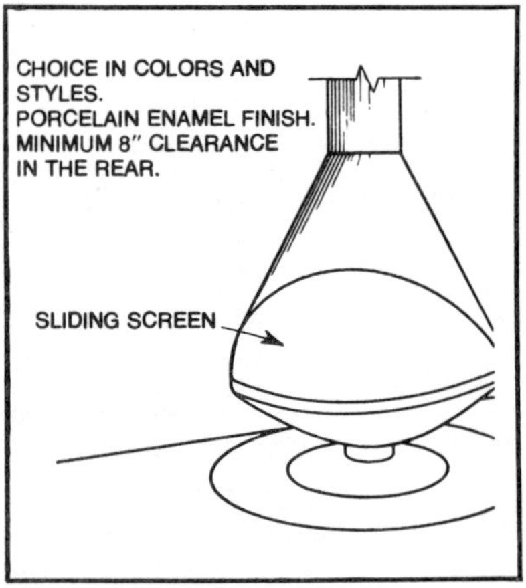

Fig. 5-7. Prefabricated fireplace and chimney units are available to offer the homeowner a safe and quick way to have a fireplace.

fireplace near any wood sheathed buildings or other objects that may cause a fire hazard.

Always try to locate the fireplace so that the front of the firebox faces the prevailing winds. This not only will help the fireplace to draw better, but will also allow the smoke to be blown away from the cook. Furthermore, a level area should be chosen for the fireplace as well as surrounding serving areas. Once these items have been decided upon and approved by the fire marshal or building inspector, the construction of the fireplace can begin.

Construction

Obviously, the foundation is the first step in building a fireplace or almost any other outdoor structure. The type, size, and material used in a fireplace foundation is governed largely by the type of soil, the climate in the area, and the amount of subsurface water. Loose soils like sand and loam are poor weight bearers. They will also settle, swell, or wash away when wet. All soils are subject to heave and buckle if freezing conditions prevail for reasonably long periods of time. The worst soil condition is fresh fill of any type. If the location you selected for building your fireplace has fill dirt, either rent a soil compact or else have it tamped with air-operated equipment.

In some areas the foundation must be designed to allow subsurface water to drain. This can be taken care of by placing a layer of gravel 5 to 8 inches thick below the foundation. To keep water from damming up behind the foundation, like on hillside locations, the gravel layer should be extended on the uphill side of the foundation. This allows the water to move around the foundation and drain off.

When heavy frost conditions can be expected, the foundation should extend down below the frost line. This allows the footing to be placed in solid ground that will not be moved by the frost conditions. The chances for frost conditions to cause tilt or creep of the foundation are lessened.

In areas where the climate is mild the year round, the foundation can often be constructed of bricks, cement blocks, or stones laid directly in a shallow trench 6 or 7 inches deep. The spaces between the bricks, stones, etc., are then filled with sand, dirt, or mortar mixed as follows: part cement, ¼ part lime and 3 parts sand. The best foundation for any fireplace is a concrete slab.

The best types of concrete slabs consist of concrete poured over wire mesh or reinforcing rods. The reinforcing gives the slab greater resistance to uneven loads caused by shifting soil. Reinforcing mesh usually has a 6 × 6-inch mesh or a 3 × 6-inch mesh. Reinforcing rods are usually ½ inch in diameter and must be placed in the concrete to form a crosshatch pattern. Regardless of the type of reinforcing, it should be placed approximately halfway between the top and bottom of the slab. The concrete should be a 1:2:3 mix, and the slab should extend down below the frost line and be at least 3 inches wider and longer than the fireplace itself.

The firebox is obviously that portion of the fireplace where the burning takes place. Because of the high temperatures encountered in this area, the firebox must be either metal or firebricks that will withstand high temperatures. Ordinarily bricks are too porous, and they will often break when heated due to differential expansion.

The construction of the firebox is also limited by the size of the grill. Before constructing the firebox, determine the grill size. Then build the firebox to support the grill. Once the grill size has been determined, the material must be selected. Metal fireboxes can be made at home if you have access to an electric welder. If not, the welding can be done at a local welding shop. Metal fireboxes can also be purchased from barbecue fixture manufacturers.

Fire-hardened bricks are another possibility for the firebox material. These bricks are as similar in dimensions and materials as

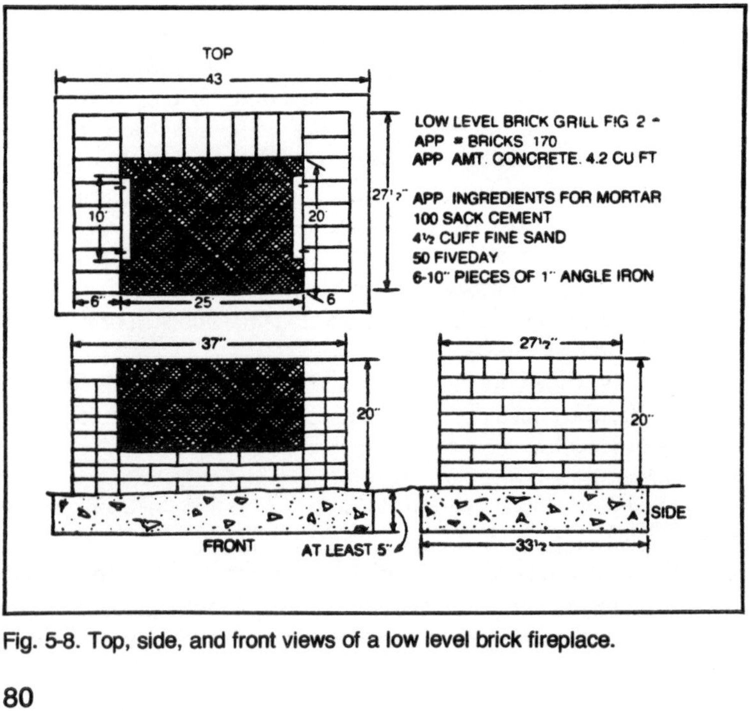

Fig. 5-8. Top, side, and front views of a low level brick fireplace.

common bricks, the only difference being that the fire-hardened bricks are baked longer. To tell the difference, hit the brick with a hammer; if a sharp clear sound is generated, the brick is hard-burned. If the sound is a dull thud, similar to striking a block of wood, the brick is not hard-burned. Besides using the fire-hardened bricks for the firebox, they are also good for use throughout the fireplace construction.

Firebricks are materials suitable for constructing the firebox. The bricks are made of a special clay that resists the high temperatures than can break common building bricks. Some important factors should be considered before deciding upon firebricks. They lack the ability to withstand moisture and cold well. The dimensions of firebricks are different from common building bricks, and they are more expensive. The standard firebrick measures 9 × 4½ × 2½ inches. If the firebricks are to be used in a fireplace design, they should be arranged so their size will fit in with the common brick. Also, due to their cost, arrange them so that a minimum of firebricks will be needed.

If bricks are to be used in the firebox construction, the mortar must be specially mixed to resist high temperatures. Heat-resistant mortar is mixed in a ratio of 1 part cement to 4½ parts fire sand to ½ part fireclay. These amounts can be measured by the shovelful, but make certain that you mix the dry ingredients well before adding water. Enough water should be added so the mortar falls from the mixing tool without sticking. In the firebox, mortar joints should be 1/16 to ¼ inch. This joint is very important since mortar does not hold heat well. If the mortar joints are large, a great amount of heat is lost to the fireplace bricks, stones, or blocks. This can cause cracking of the mortar or even the masonry material.

CONCRETE

Concrete is usually measured in cubic yards. To determine the amount of concrete needed for a particular project, multiply the height, width, and depth (in feet) of the area. Then divide by 27 cubic feet to obtain the volume in yards. It might be best, depending on the size of the foundation, to determine the actual amount of concrete to order after the foundation trench has been dug. In any case, the amount of concrete can be determined by the number of cubic feet of the foundation.

If the volume of the foundation is equal to or less than 1 cubic yard, ready to mix concrete is available that only needs water added. Any amount greater than 1 cubic yard should be ordered from

Table 5-2. Amount of Ingredients to Use for Ready to Mix Concrete.

cubic ft. of concrete	1	3	6	9	18	27
individual ingredients						
Bags of cement	1/5	3/5	1⅓	2	4	6
Cubic Feet Sand	½	1½	3¼	4¾	9½	14½
Cubic Feet Gravel	4/5	2½	5	7⅔	15⅓	23

a ready-mix concrete plant. If you want to save a little money on the ready to mix concrete, mix your own from separate ingredients. Table 5-2 should help in figuring the amount of separate ingredients.

To mix the ready to mix concrete, just add water—the amount specified in the instructions on the bag. On the other hand, mixing the individual ingredients is a little more complicated. In the latter case, the concrete should be mixed to a ratio of 1 part cement to 2½ parts of sand to 4 parts gravel (1:2½:4). Start by mixing a test batch of 1 shovelful of cement and 2½ shovelfuls of sand. Mix well until all of the color is uniform. At this point, add 4 shovelfuls of gravel and mix until the gravel is evenly distributed. Add 3 quarts of water and mix until all of the dry mix is dissolved. If the mixture is too soupy, add more sand or gravel. If the mixture is too stiff, add more water to the batch.

When the concrete is mixed or delivered, pour it into the foundation trench and tamp each batch within 15-30 minutes after pouring. When the foundation is half filled, lace reinforcing wire or rods on top of the pour. The reinforcing rods should be laid in such a way that they form a mesh. Then finish pouring the concrete, tamping it around the rods to mesh so they are completely encased. When the concrete fills or the top of the foundation trench, scrape the surface level with the flat edge of a board long enough to lay across the foundation. After smoothing, check the surface with a level and trowel out any irregularities.

BRICKS,BLOCKS, AND STONE

Brick dimensions are usually 8¼ × 3-⅞ × 2½ inches wide. These dimensions are not exact and can vary as much as ½ inch. Mortar joints should be ½ inch.

To estimate the number of bricks needed for a particular project, 100 square feet of wall will require approximately 617 bricks and 8 cubic feet of mortar. You can figure about 6.17 bricks per square feet for a 4-inch wall. For an 8-inch wall, using two courses of brick, figure 12.33 bricks per square foot.

Assume that the area of three walls for a small fireplace totals 18 square feet. The approximate amount of bricks needed is found by multiplying 18 by 6.17, which equals about 112 bricks.

To estimate the amount of mortar required, the ratio of 1 part cement to 4-4½ cubic feet of sand to one 50-pound sack of fireclay will provide enough mortar for 250 bricks. You would need about half this amount of mortar, since there are only 112 bricks.

Using 20 pounds as being approximately equal to 1 shovelful, you can calculate that enough mortar for 50 bricks can be made by mixing 1 shovelful of cement to 4½ shovelfuls of sand to ½ shovelful of fireclay.

Concrete or building blocks are usually 8 × 8 × 16 inches (with mortar joints). The number of blocks needed for a project can be estimated by calculating the square footage of the walls and chimney of the fireplace, and then multiplying the area by 1.125. Mortar for the blocks is the same as for bricks. The joint should be as close to ½ inch as possible. For every 100 square feet of blocks, you will need approximately 2.6 cubic feet of mortar.

Stone is more difficult to estimate due to its irregular shape. To roughly estimate the amount of stone needed, calculate the cubic footage of space to be filled. Then take this information to a stone supplier for his estimate. He will be able to tell just how much stone will be needed as well as how much mortar to use. Incidentally, the mortar for stone is usually richer in cement and fireclay than that used for bricks and building blocks. Stone mortar should be mixed at a ratio of 1 part cement to 3 parts clean sand to ½ part fireclay.

With the information provided thus far, you should be able to construct any of the fireplace designs to follow. These plans are basic ideas and can be easily modified to suit your particular need or to fit any backyard setting.

FIREPIT

The *firepit* in Fig. 5-9 is simple and inexpensive to build. The tools needed include a shovel, pick or digging iron, and trowel. You will need about 19 cubic feet of blue stone, 7 cubic feet of pea gravel, four angle iron 28 inches long, and 20 to 30, 8 to 10-inch diameter stones. If you mortar the joints, mortar will also be needed.

Start by digging a circular pit, 24 inches in diameter and 29 inches deep. Arrange the stones around the edge of the circle. Then fill the pit to within 5-7 inches from the top with bluestone. Fill the remainder with the pea gravel.

The diameter of the circle measures 12 inches to each side of the center. Drive a piece of angle iron into the gravel at each of these points and repeat this operation perpendicular to the line formed by the first two pieces of angle iron. These are the supports for the grill and should be driven into the ground, leaving only 10 inches above the pea gravel. With the supports in the position described, they should form a square 16 inches on each side to take a 25-inch barbecue grill. Fasten the grill to the four posts with stove bolts. You can then use either wood or charcoal as a fuel for your cookout.

RUSTIC GRILL

For those persons living in areas with lots of native stone above ground, the fireplace shown in Fig. 5-10 can be quickly constructed in one day. Once the concrete cures, you'll be ready for that long awaited cookout.

This little fireplace is merely a grill with native rock on sides and one end. The rock selected should be a type that will withstand heat. The sides facing the firebox and top should be relatively flat. This means solid nonporous stone.

Begin construction of this fireplace by digging a hole in the ground at the building site approximately 3 feet 2 inches wide by 3 feet 6 inches in length. The depth should be below the usual frost line for the area. Concrete footing may then be poured, or concrete blocks may be laid in the pit with their smooth sides up.

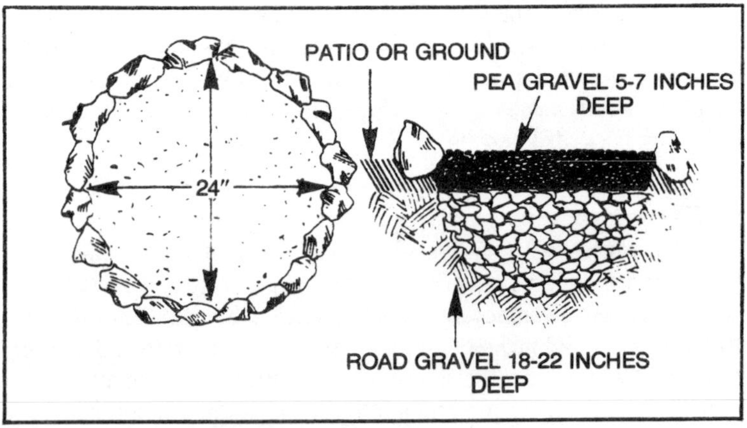

Fig. 5-9. A firepit is an inexpensive way to cook outdoors.

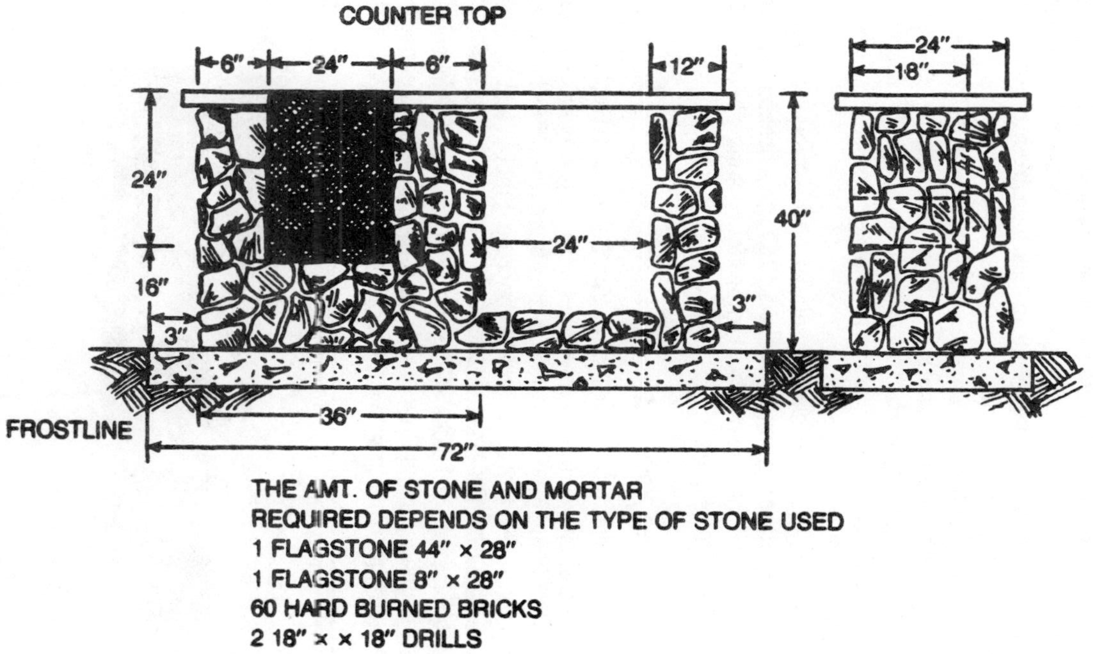

Fig. 5-10. This stone fireplace is a good project for the homeowner who has building stones on his property.

The metal grill is constructed with 13 iron bars, ½ × 1 × 15 inches, welded on both sides of a 1¼ × 1¼ × ¼-inch angle iron, 2 inches on center. A ¼ × 6 × 15-inch end plate is then welded on the back end, and ¼ × 24-inch side plates are welded to the sides. Braces and legs consist of 1¼ × 1¼ × 1¼-inch angle irons, all welded together. Note that the grill top is 15 × 24 inches outside dimensions and approximately 10 inches high from the concrete hearth to the top of the grill.

Since three sides of the firebox are enclosed with iron plates, the stones are mainly for decoration. They also help to secure the firebox. None of the stones or concrete blocks should be cemented together in this design. If you can find three large stones, one for each side, so much the better.

Many campers have used fireplaces of this type constructed only of stones placed around three sides to form a firebox. An oven grill from an old cooking range will suffice for cooking when the grill is laid across the stones.

LOW LEVEL BRICK FIREPLACE

The low level brick fireplace in Fig. 5-8 is a relatively simple project to build if you have some experience at laying bricks. Start by digging a foundation 43 × 33½ inches by at least 5 inches deep. Mix approximately 4.2 cubic feet of concrete. Smooth and level the concrete as explained previously, and let the concrete cure.

Begin laying the bricks at the corners (Fig. 5-11). Measure in from the outside edges 3 inches. Where the lines intersect at the corners is approximately the position of the brick corners of the

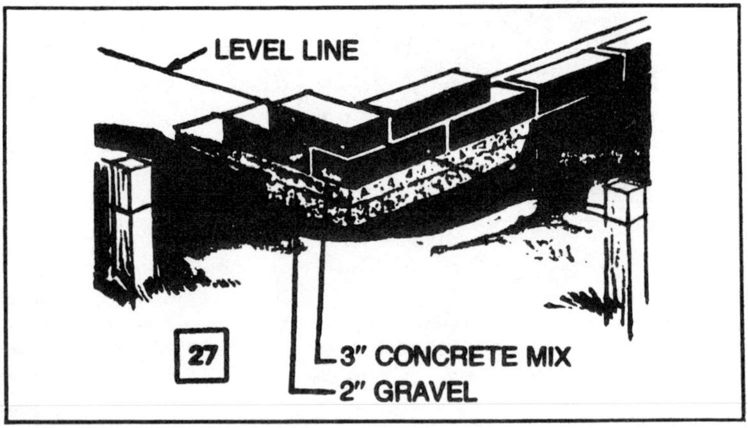

Fig. 5-11. Method of laying brick corners.

fireplace. Check to make sure by measuring between the marks. The measurements should be 37 × 27½ inches. If the measurements don't come out to these figures, use the following method.

Start on one of the corners on the longest side and mark a line to the other corner marks (again on the longest side). Measure the line from one of the corners and mark 37 inches. Take a square and mark a line to one of the corners on the short side. This line should be 27¼ inches. Using the square, mark the other two edges.

Lay two or three rows of bricks at right angles to each other at each of the two back corners. Lay the outside bricks first and then the inside bricks. After the bricks are in place, brush all excess mortar off and allow the mortar to cure. Then the grill supports can be installed. These supports are made from 1-inch angle iron and should be mounted to form an "L" bracket for the grill to rest on. The bottom bracket holds the fire grill, while the upper two brackets support the cooking grill at two levels from the cooking fire.

Use 1½ × 5/16 inch lag screws to bolt the angle iron into position. The fire grill should be mounted 6 inches above the base of the firebox, while the first set of brackets for the food grill should be mounted 5 inches above the fire grill and the second set 5 inches higher.

The grill can be made by welding 17 iron bars, ½ × 17 inches, at equal intervals to two ½ × 24½-inch iron bars. The fire grill can be made from 17 iron bars, ½ × ¾ × 17 inches welded at equal intervals to two pieces of 1-inch angle iron 24½ inches in length. The fire grill can be permanently mounted on the lowest bracket or can be left loose.

COUNTERTOP GRILL

The fireplace is designed so that the grill is counter height—eliminating the need to bend over while cooking. This design also allows for a counter and storage space. It can easily be modified to include doors on the storage side, or a prebuilt warming oven can be incorporated into the unit.

You will need approximately 60 fire-hardened bricks for the firebox, one 28 × 44-inch flagstone, and another 8 × 28 inches in size for the countertop. Also required are two 18 × 18-inch grills, six 18-inch pieces of 1-inch angle iron, and enough concrete to pour a foundation 6½ × 2½ feet, deep enough to reach below the frost line in your area. You will also need reinforcing wire for the foundation, and enough stone (or brick if you desire) to build the walls and chimney.

Prepare the trench and foundation as described earlier. When the concrete has cured, mark the boundaries of the fireplace on the slab. Begin laying the brick or stone at the corners first. Then work up to the top filling in between the corners—leaving space for the firebox. Fill all cracks around the stone with mortar. Make the mortar joints as thin as possible. Use small pieces of rock to fill in large spaces between rocks. If a stone needs to be moved after it has already been set in mortar, remove it completely from the mortar. Then remove all of the mortar before starting again.

Before the mortar joints set, take a stick or piece of small diameter pipe and scrape mortar from around the rocks to give the joints an inset. Overlap all of the vertical joints with longer rocks to strengthen the wall. Thoroughly brush off all the excess mortar before it dries.

When the stones have set, spread a layer of brick mortar over the bottom of the firebox inset. Lay the bricks for the bottom of the firebox on this layer of mortar, making sure that they are level. Keep this boundary layer of brick mortar as thin as possible. After completing the bottom layer of bricks, continue by completing the walls.

The firebricks should be laid so there is 18 inches between the two sides. When the firebox is complete, fasten the angle iron brackets to the walls at 8-inch intervals, with 1½ × 5/16-inch bolts. Place the flagstone countertops on the fireplace so that the overhang is 2 inches on all sides, except where they come in contact with the firebox. The space between the two pieces of flagstone behind the firebox should be filled with smoothed brick mortar. The space under the countertop can be used to store wood, charcoal, and similar items.

Another popular outdoor fireplace is shown in Fig. 5-12. This design can be completed in a weekend by digging and pouring the foundation on Saturday, letting it cure overnight, and then laying the bricks on Sunday afternoon. All of the necessary construction details are provided in Fig. 5-12.

ESTIMATING MATERIALS FOR FIREPLACES

Most fireplaces are of irregular shape. Brick and other materials are often estimated by the cubic foot method. This method involves obtaining the total cubic feet of brickwork required and then deducting the openings for flues and ashpits.

Figure 5-13 shows a fireplace requiring face bricks on the outside. One method of estimating the number of bricks required is

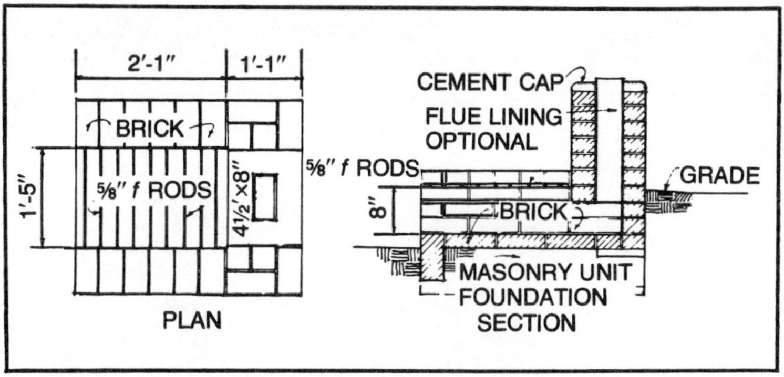

Fig. 5-12. A very popular design for an outdoor fireplace.

as follows. The portion under the first floor is 6 feet × 2 feet, or 12 square feet as shown in Section A-A. The openings illustrated by Section A-A, as might be found on a basement plan, are approximately 1 foot in width. The combined length is about 4 feet. (This is the entire length of the four openings combined.) The section area of the opening would be 4 × 1, or 4 square feet. The net area of brickwork is 12 – 4 = 8 square feet. The height is 8 feet. The total cubic feet of brickwork would be 8 × 8, or 64 cubic feet. Assuming 21 bricks per cubic foot, you have 64 × 21, or 1344 common bricks. The arch carrying the hearth is about 2 feet × 6 feet, or 12 square feet in area. At seven bricks per square foot, it would require 7 × 12, or 84 bricks.

Notice that this would really include part of the face bricks above the grade. For ordinary estimating, this need not be considered.

Sections B-B, C-C, and D-D drawn above the first floor show how many bricks would be required per course. In Section B-B you see that the outer portion requires face bricks. Likewise the face of the fireplace calls for face bricks. The only portions that require common bricks are the filling in bricks and the bricks on the end which would probably be plastered over.

You would have approximately 10 to 12 bricks per course in Section B-B. In Section C-C you would have about nine bricks per course. In Section D-D you also have approximately nine bricks per course. As an approximate average, you may take 10 bricks per course in the height from the first floor to the top of the roof. This would make 10 × 4.36, or 44 bricks per foot of height. The distance from the first floor to the top of the roof is 25 feet. This gives you a total of 25 × 44, or 1100 common bricks.

Above the roof the entire chimney would be built of face bricks, except perhaps the portion between the two flues. If you wish to include this in your estimate of common bricks, you may assume two bricks per course or 2 × 4.36 = 8.72, or nine bricks per foot in height. The chimney extends 6 feet above the roof. This portion, therefore, would require 6 × 9, or 54 common bricks.

Adding up the estimate of common bricks:

For the portion below the first floor	1344 bricks
For the arch supporting the hearth	84 bricks
For the first floor to top of the roof	1100 bricks
For the part above the roof	54 bricks
Total common bricks	2582 bricks

Adding for breakage, you may assume that 2600 bricks would be required.

In estimating the firebricks required for the lining of the fireplace, you may either figure the square feet of area or calculate them

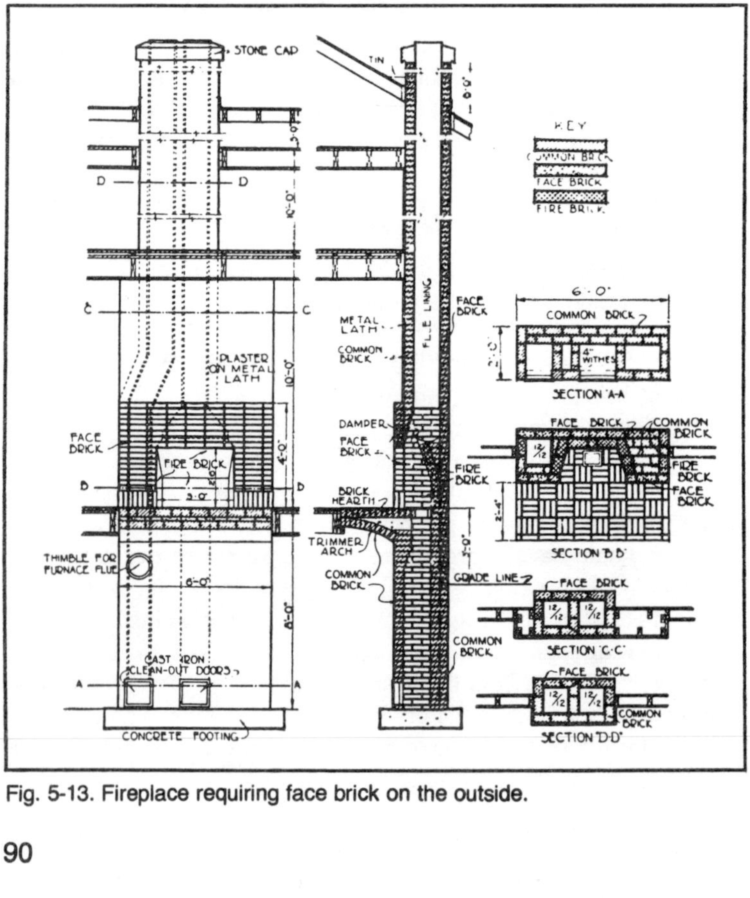

Fig. 5-13. Fireplace requiring face brick on the outside.

by the courses. Section B-B shows that between six and seven firebricks are required per course. The fireplace opening requiring firebricks is approximately 2 feet 6 inches high. At seven bricks per course, you would have 7 × 4.36 = 30.52, say 31 bricks per foot, or 31 × 2½ = 77.5, say 78 firebricks for the fireplace lining.

The bricks required for the hearth may be found by the square foot method. The main portion of the hearth is 6 feet × 2-1/3 feet = 14 square feet. The portion in the fireplace is 2½ feet × 1½ feet = 3-¾ square feet. The total square feet of hearth is 14 + 3-¾ = 17-¾, or 18 square feet. At seven bricks per square foot, you would require 7 × 18, or 126 bricks for the hearth.

The bricks for the front fo the fireplace may also be found by the square foot method. The front measures 6 feet × 4 feet or 24 square feet. From this you must deduct the fireplace opening. This is equal to 3 feet × 2½ feet, or 7½ square feet. The total area requiring face bricks for the face of the fireplace is 24 less 7½ = 16½, say 17 square feet. This would require 17 × 7, or 119 face bricks.

The exposed area of the chimney on the outside of the building also requires face bricks. The number of face bricks required per course is found in Section B-B, Section C-C, and Section D-D of Fig. 5-13. In Section B-B, you find that it requires 10 bricks per course. This would make 10 × 4.36 = 43.6, or 44 bricks per foot of height. About 10 feet of the chimney requires this number of face bricks. That is, the chimney is this width up to 10 feet above grade. This portion would require 10 × 44 = 440 face bricks.

From this point to the roof line is about 18 feet. This part would require face bricks as shown in Section C-C and Section D-D. Six bricks are required for each course or 6 × 4.36 = 26.16, say 26½ bricks per foot in height. For 18 feet, it requires 18 × 26½ = 477 face bricks.

The part above the roof is 6 feet and is built the same as shown in Section D-D. Face bricks are required for all four sides of the chimney. Therefore, 12 bricks per course would be needed, or 12 × 4.36 = 52.32, say 53 bricks per foot in height. This will make 53 × 6 = 318 face bricks. The total number of face bricks required for the outside of the chimney would be 440 + 477 + 318 = 1235.

Summarizing the different kinds of brick required, you have the following:

Common bricks	2600
Firebricks for lining	78
Face bricks for hearth	126

Face bricks for face of fireplace	119
Face bricks for outside of chimney	1235

Using the standard size firebricks, 2½ × 4½ × 9 inches, you need 400 to 600 pounds of fireclay to lay 1,000 straight bricks. You also have the following:

1 square foot of 2½-inch thick wall requires	3.58 bricks
1 square foot of 4½-inch wall requires	6.31 bricks
1 square foot of 9-inch wall requires about	14 bricks
1 square foot of 13½-inch wall requires about	21 bricks
1 cubic foot of firebricks requires	17 bricks
1 cubic foot of firebricking weighs	125 to 140 pounds

Chapter 6
Practical Chimney and Fireplace Designs

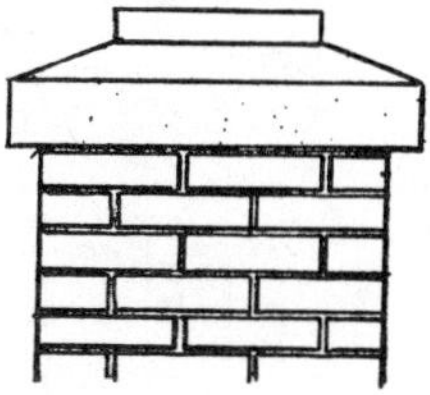

Proper construction of chimneys and fireplaces is essential for safe, efficient operation. They must be designed and built by persons experienced in that type of work. The designs of chimneys and fireplaces described in this chapter are those which have actually been built and are in use. While any one of these designs will provide you with safe, adequate use, your main concern will be to select a design that best fits your needs. If you plan to heat only with a woodburning stove, you will likely want to build only a chimney and not continue with a fireplace in your home. If you want a fireplace only for appearance, you'll want to forget about a chimney suitable only for woodburning stoves, etc.

Read through the description of the various designs to get an overview of the types available. Then decide which design best fits in with the scheme of your home and your intended use. Try to visualize just how the chimney or fireplace will look in your home. Making sketches will help immensely. You should then make detailed working drawings to see if the selected design will fit into the allotted space. The drawings act as guides during the construction. Before starting the construction, however, you will want to survey the structural conditions of your home. See what wooden members will have to be cut away and how to replace the weakened structure during construction of the chimney or fireplace. You can't go cutting away at structural members without careful planning. Each member is there for a purpose, and the removal of just one could cause a collapse of part of your home.

Make sure you know what you are doing before attempting any major alteration on the structural members of your home. If you

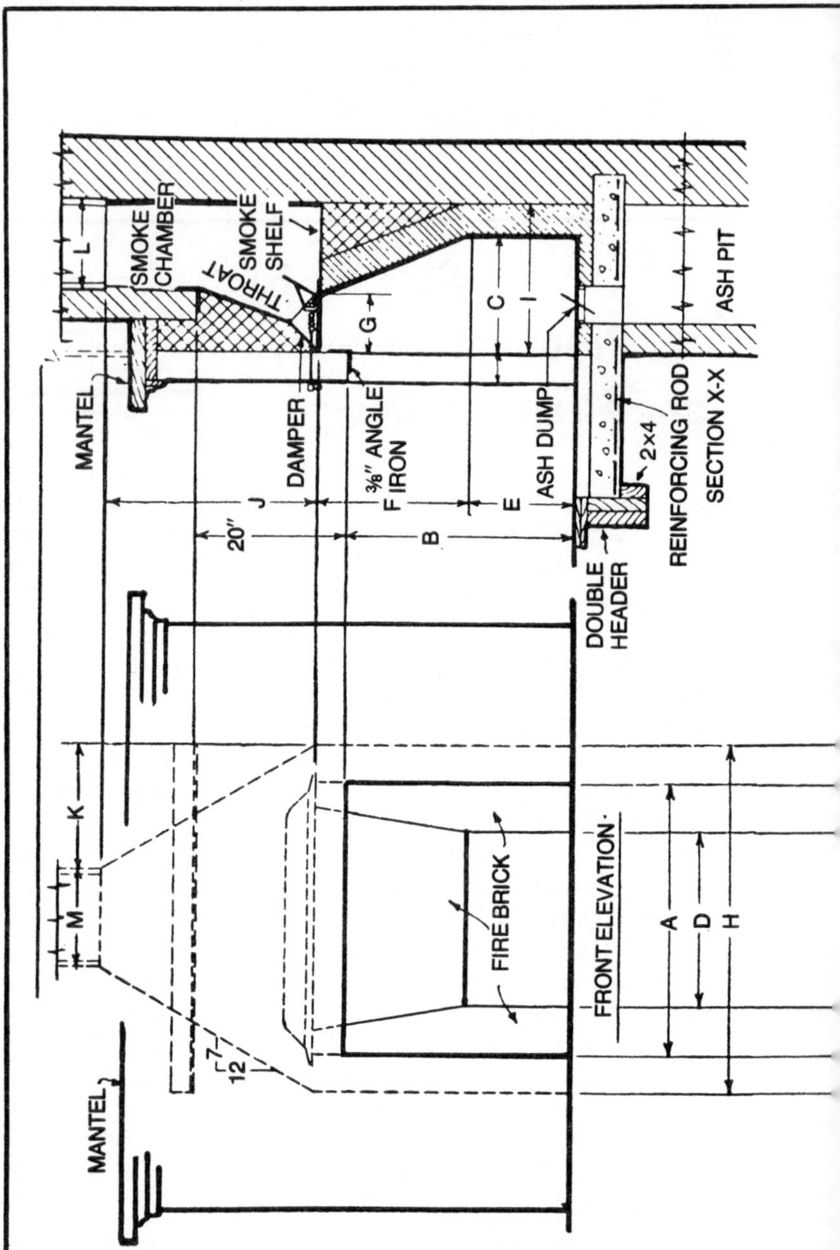
MANTEL
SMOKE CHAMBER
THROAT
SMOKE SHELF
L
C
I
G
DAMPER
3/8" ANGLE IRON
ASH DUMP
ASH PIT
J
F
E
20"
B
2×4
REINFORCING ROD
SECTION X-X
DOUBLE HEADER
M
K
FIRE BRICK
FRONT ELEVATION
A
D
H
7
12
MANTEL

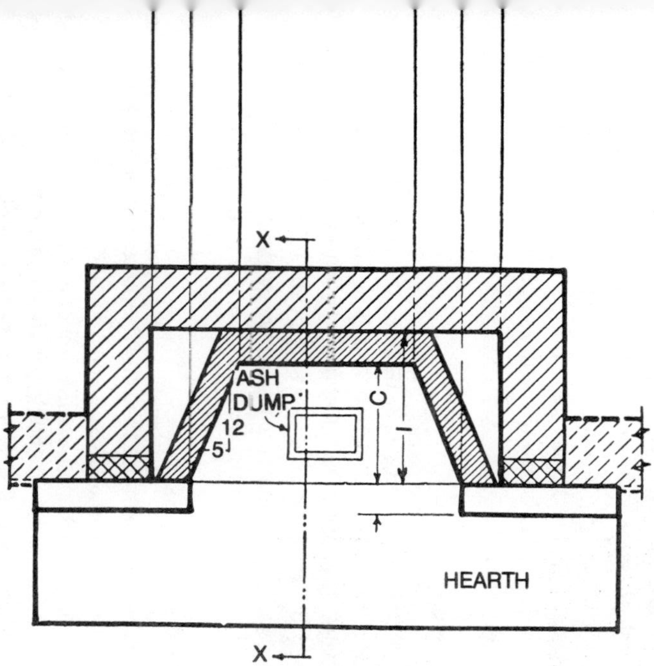

FIREPLACE DIMENSIONS

Finished Fireplace Opening								Rough Brickwork					
Width	Height	Depth	Back	Vertical Back Wall	Sloped Back Wall	Throat	Width		Smoke Chamber	Slope of smoke chamber	Std Rectangular Flue Lining outside dimension		Std Round Flue Lining inside diameter
A	B	C	D	E	F	G	H	I	J	K	L	M	ϕ
in.	in.	in.	in.	in.	in.	in.	in.	in.	in.	in.	in.	in.	
24	28	16	11	14	18	8	37	20	24	14	8½×8½		
28	28	16	15	14	18	8	42	20	25	14⅛	8½×13		10
30	30	16	17	14	20	8	42	20	25	14⅛	8½×13		10
34	30	16	21	14	20	8	46	20	28	16½	8½×13		12
36	30	16	23	14	20	8	46	20	28	16½	13×13		12
40	30	16	27	14	20	8	50	20	32	18½	13×13		12
42	30	16	29	14	20	8	54	20	35	20½	13×13		15
48	33	18	33	14	23	8	59	72	40	23	13×13		15
60	39	22	42	14	29	1	71	26	45	26½	18×18		18
72	40	22	54	14	30	12	83	26	56	32½	18×18		18

Fig. 6-1, Elevation and section views of a typical fireplace.

aren't sure, seek professional advice. If major alterations are necessary, many homeowners find it more convenient (and often cheaper) to hire a responsible contractor to do the work.

The plans to follow are complete working drawings for various chimney and fireplace designs. The details of construction are complete. You need to know certain terms and symbols used in construction drawings.

ELEVATION AND SECTION VIEWS OF A FIREPLACE

The elevation and section views of a typical fireplace are shown in Fig. 6-1. The two left-hand columns give the more common sizes of fireplace openings. By selecting any size opening and reading from left to right, the correct dimensions for the various sections of a particular fireplace can be obtained. If you choose to build a fireplace with an opening of 30 inches wide by 30 inches high, this dimension is found in the two left-hand columns. Reading from left to right, you see that the depth of the fireplace should be 16 inches: the back, 17 inches; the vertical back wall, 14 inches; sloped back wall, 20 inches; and the throat, 8 inches. By reading further, you see that the rough brickwork will be 42 inches wide by 20 inches deep; the smoke chamber, 25 inches; the slope of the smoke chamber, 14½ inches, and the rectangular flue lining, 8½ × 13 inches. Should round flue lining be used, the fireplace will require a diameter of 10 inches.

TWO FIREPLACE DESIGNS

Another design is shown in Fig. 6-2. Besides dimensions, the drawings show details of the foundation and how the concrete hearth joins the wood framing members. The design in Fig. 6-3 provides an opening for an ash dump which should end either in the basement of the house, or to the outside just above grade level. The broad template used as a guide for the masonry work is also shown.

A DESIGN WITH TWO FIREPLACES

It may be desirable to have more than one fireplace, one above the other, and each fireplace should have a separate flue. If you also plan to use a furnace that burns combustible fuel or a wood stove in conjunction with the fireplaces, you should provide a separate flue. One such design appears in Fig. 6-4. Here you see two fireplaces at two different levels, one above the other. Notice how the flues are offset to avoid obstructions and to line up neatly in the chimney. In the front elevation view, the flue to the left of the fireplaces is for a

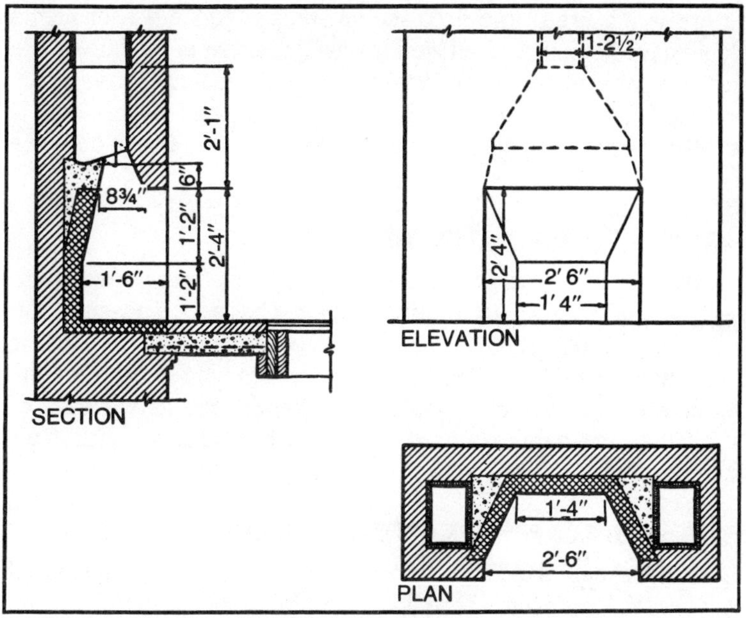

Fig. 6-2. Another design for an open front standard fireplace.

furnace or wood stove. While shown on the first floor, it could have an opening on the second floor.

When using the arrangement in Fig. 6-4, it's best to have the flue opening for the furnace or wood stove on the opposite side from the fireplace if this is practical. Furthermore, it's best to have a partition between the fireplace and the area in which the furnace or wood stove will be. If the chimney is designed and built properly, you should have little trouble with short circuiting between the fireplace and woodburning stove, but I still remember some agonizing moments in a mountain cabin my family and I lived in some years ago.

The only source of heat was a fireplace and a woodburning range in the kitchen, which was merely an extension of the living room. Although the cabin also had an electric range for cooking, it was necessary to light the wood range on very cold mornings to supplement the fireplace. The only problem was that the flue in the fireplace created more draw and draft. It would continually suck smoke from the firebox of the range into the room. We finally had to discontinue use of the fireplace altogether—sealing up the opening airtight—and rely solely on the woodburning range in the kitchen. The wood range alone provided adequate heat for the little cabin, but a roaring fire in the fireplace on some of those winter nights would have been nice. Had we increased the size of the flue on the chimney where the range was attached to equalize it with the fireplace chimney, we probably wouldn't have had the problem. I still think a partition between the two heaters would have done the trick.

ARRANGEMENT FOR TWO OR MORE FLUES

When two or more flues are to be installed in the same chimney, an arrangement such as Fig. 6-5 should be used. If possible, the design on the left (with a brick partition between the two flues) is the preferred method. If sufficient concrete is poured between the flue like the arrangement on the right, however, this will suffice. Just make certain there are no leaks in the flues so as to shortcircuit each other.

INSTALLATION OF PREFABRICATED FIREPLACES

I have discussed only those fireplaces that are considered conventional in design,—the open front types that project heat out only in one direction. While the open front design is the most common, there are other types that you may want to consider. Prefabricated, self-contained freestanding fireplaces are available in various designs. When installing a prefabricated freestanding

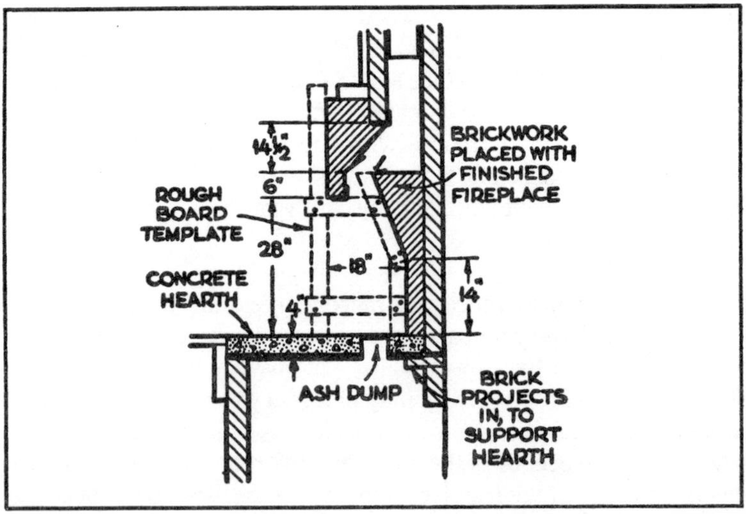

Fig. 6-3. This design provides an opening for an ash dump which should end either in the basement or to the outside just above grade level.

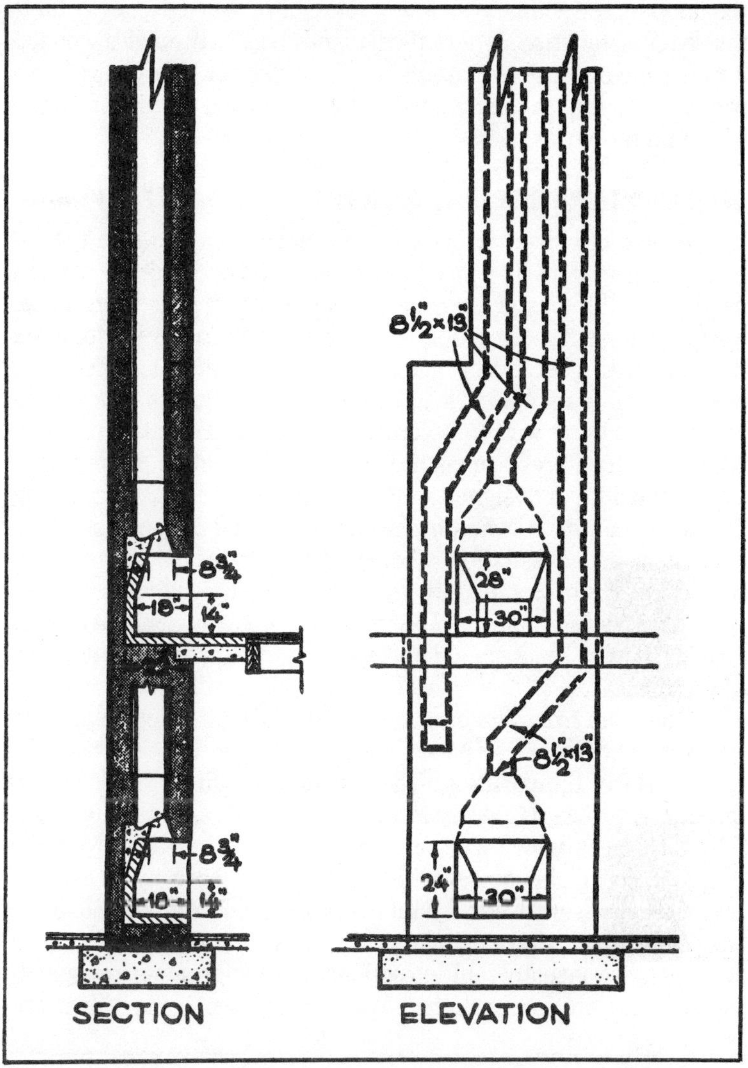

Fig. 6-4. This design not only provides for two fireplaces, but an additional flue is added to the chimney for a fuel-burning furnace.

fireplace, it should be set on an incombustible base. This can be a concrete hearth or a prefabricated base especially designed for such fireplaces. When these fireplaces are used, a metal-lined and insulated exposed flue is normally installed which runs to an existing chimney, rises through the ceiling of the house, or runs through the wall to the exterior. The most care must be taken when installing the metal-lined chimneys. Many fires have started due to their improper installation. Where the chimney passes through a wooded framing member, floor, and the like, be certain that you use approved fittings as recommended by the NFPA and the manufacturer of the unit.

OPEN FRONT AND SIDE, AND OPEN FRONT AND BACK FIREPLACES

A type of built-in fireplace is the open front and side type—throwing heat in two or more directions. The open corner may be cantilevered or post-supported depending upon your preference. The primary visual direction of this type of fireplace is along the diagonal, with a secondary one perpendicular to the long face. They are especially useful when a fireplace must be built in the corner of a room, but on one wall (not across two walls). The main problem with multifaced fireplaces is the problem of smoking. Even though they are built and designed well, they are still subject to smoking due to certain draft conditions which may exist in the room. The placement and orientation of these types of fireplaces are very important. Each specific situation should be carefully considered regarding potential draft problems. Many homeowners have been forced to have a fan installed in their flues to insure proper drawing of the fireplaces.

The open front and back types of fireplaces are very popular when located within a space dividing partition and thus used on either side of the partition. These fireplaces project heat in two directions—towards two spaces.

Fireplaces with three or four open sides are used to some extent in certain house designs. The open corners may be cantilevered or post-supported. Metal fireplace hoods and flues may be suspended from above, similar to freestanding fireplaces. These types may be visually multidirectional or omnidirectional depending on form and proportions. Again, these types may have the

Fig. 6-5. When two or more flues are installed in the same chimney, this arrangement should be used.

problem of smoking if not orientated correctly as to drafts in the room.

Drawings of some of the fireplaces just described are shown in Fig. 6-6. A better knowledge of their use in your own home can be determined by examining model homes with the various designs.

Prefabricated fireplaces are a good choice when adding a fireplace to an existing home. Most have metal exteriors with heat-resistant insulating liners that may be used in most standard buildings with a zero clearance requirement between the heating unit and adjacent surfaces. Standard components available include a built-in damper, forced air blower, chimney starter sections, prefabricated flue sections, and chimney houring tops. Masonry is normally built up around the liner once it has been properly installed.

CHIMNEY FOR A FURNACE OR WOOD STOVE

If you only want to provide a way to carry the smoke from a furnace or woodburning stove, then you might prefer to build only a chimney and forget about a fireplace. In doing so, make sure the minimum wall thickness is 4 inches. The overall dimensions depend on the size of the flue tile used. An 8½ × 18-inch flue should be sufficient for most applications.

The footing of the chimney should be at least 12 inches deep and extend 6 inches beyond the chimney on all sides. If you are building without a basement and the house is on masonry piers, it is important to have the footing extend well below the frost line. Keep the masonry at least 2 inches away from all the house framework.

Provide a cleanout door located below the entrance of the thimble if practical. Flue tile should extend at least 8 inches below the thimble. Many builders carry it all the way down to the base of the chimney or at least to the top of the cleanout door.

The following materials were used to construct one chimney to which a woodburning stove was attached. The list should prove useful as a guide for you to estimate your own materials.

64	chimney blocks	4 × 8 × 12 inches
9	clay flue tile sections	
1	concrete chimney thimble	
1	cleanout door	
1.5	cu. ft. cement mortar	
	copper or tin flashing	

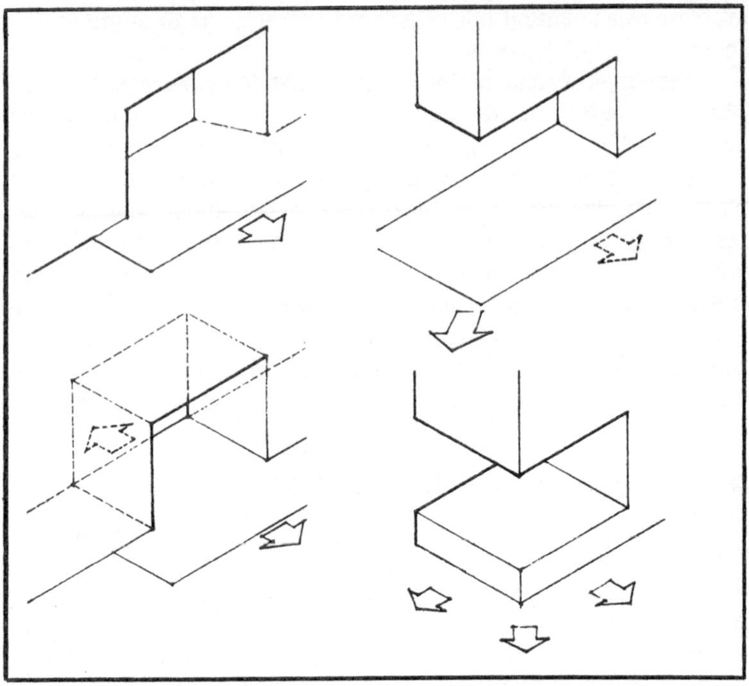

Fig. 6-6. Several types of fireplaces showing the direction of heat.

When fireplaces or chimneys are built in a building with a concrete floor, the work is somewhat simplified. There are still factors to consider.

First, a special footing is poured for the chimney or fireplace that is completely independent of the concrete slab. Sufficient room is also left for the hearth, although it need not be as deep as the footings for the chimney as it has not nearly as much weight. Still, it should be independent from the concrete floor.

Once the footing has been finished, the construction of the rough chimney or fireplace can begin right on top of it. The work goes in the same manner as for a conventional fireplace.

Chapter 7
Curing Smoky Stoves and Fireplaces

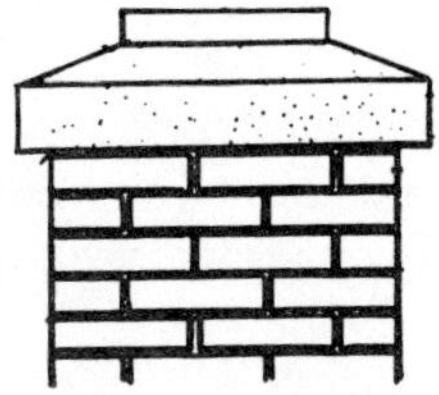

One of the most annoying problems with fireplaces and woodburning stoves is when they smoke in the wrong direction. Chimneys are designed to take advantage of the fact that hot air rises. Hot air carries the smoke and fumes away from the firebox out into the air. If there is some error in construction of the chimney, or the stove connection to it, the draft will not be generated properly. The smoke will swirl back into the stove or fireplace and eventually flow into the room. Any type of obstruction will cause the same problem.

OBSTRUCTIONS

The most obvious cause of smoky chimneys is a blockage or obstruction in the chimney, especially if it is the first time the chimney has been used at the beginning of a heating season. Certain birds build their nests in chimneys. These nests can partially or fully obstruct the rising of smoke when the fire is lit. A visual inspection, using a mirror if necessary, will let you readily detect such a problem. You can crumple up a sheet of newspaper, place it in the damper opening or chimney opening (whichever the case may be), and light it. If the paper burns easily and the smoke and flames rise upward, then the flue is not obstructed and the fault lies elsewhere. If smoke does not rise freely, chances are you have a partial or full blockage of the chimney flue. Perhaps some bricks have worked loose or some other debris is lodged in the flue.

In the case of a fireplace, open the damper fully on a sunny day and look up through it to see if you can see daylight. Due to the curvature of the fireplace throat, you probably won't be able to see up through the flue. You should be able to see daylight if your chimney is not blocked. Should an obstruction be evident, you may want to seek professional advice—like calling in a chimney sweep. You may be able to dislodge the obstruction with a long pole and remove it through the flue opening or through the throat of the fireplace.

Besides obstructing the flow of smoke and hot air, bird and animal nests can also be a hazard. If there is a buildup of creosote on the flue walls, it will normally take a sizable blaze to set it aflame. On the other hand, a small spark can easily ignite a bird's nest. This is almost certain to start a roaring chimney fire if you have creosote buildup in your chimney. You should check periodically for such nests regardless if your chimney is smoking (the wrong way) or not.

LACK OF AIR

Any type of open flame appliance needs air for it to burn properly. Most modern homes are so well insulated, and doors and windows are so well weather stripped, that it is sometimes difficult to obtain enough air to supply oxygen to the flames of fireplaces, wood stoves, or even gas and oil furnaces. This is another cause of smoky fireplaces and woodburning stoves. To test for this problem, first try cracking a window or door to see if this will help. If so, you should provide some better and permanent means of ventilation.

If your fireplace has an outside ash cleanout door, this can be cracked to allow air to enter the fireplace. Be sure to provide a screen over the opening to keep out insects and foreign matter. If no such ashpit exists, you can remove some of the bricks on the outside of the fireplace, just above the hearth level, and insert a ventilator. This will carry fresh air from the outside directly to where it is needed. In doing so, make sure you have an adjustable damper on the ventilator so that you can control the amount of primary air entering the fireplace, and that you can completely shut it during periods when the fireplace is not in use. Such an installation is not possible on inside chimneys, or when you are using a wood stove piped into a chimney flue. You might try installing a small ventilator in the outside wall of the room, equipped with a damper to control the amount of air admitted and to close completely when the stove is not in use.

The cause of a smoky stove or fireplace may be due to the way the air travels inside the room containing the heater. A door in the room may be in such a position that when it is open at a certain angle, it causes a draft to flow across the fireplace opening and drive smoke out into the room. In most cases, the draft can be diverted in another direction by changing the door opening. Hinge the door on the opposite side of the jam, or have the door swing outward instead of inward—or vice versa. If this is not possible, a fire screen installed in the fireplace will usually take care of the problem. This is merely a sheet metal shield to divert the smoke upward and the draft away from the firebox.

Many homeowners have corrected the problem of improper draft by enclosing their fireplaces with a glass fireplace screen containing small intakes for primary air at the bottom. The cast-iron stove inserts for fireplaces work even better and provide for more efficient heating.

FLUE SIZES

The size of the flue (cross sectional area) and the height of the chimney absolutely limit the size of woodburning stove or the size of the fireplace opening that can be used. If the flue opening is too small for the stove or fireplace, it will not adequately carry the hot air and smoke from the firebox. This will cause some of the smoke to enter the room or area containing the heating device.

The area of the flue opening for fireplaces must not be less than one-twelfth of the area of the fireplace opening. The chimney should be as straight as possible from the point of the fireplace or stovepipe connection to above the highest projection of the roof. The chimney should be independent, having no connection with other flues or openings, and always of the same area from top to bottom. A well jointed tile flue, preferably round, is better than a square brick flue of larger area. The joints between tiles must be tightly cemented, and inside dimensions of the tile are the size of the flue. If the chimney height is less than 22 feet, then the area of the flue should be at least one-tenth of the area of the fireplace.

A fireplace opening 42 inches wide by 30 inches high has an area of 1260 square inches. One-tenth of this figure is 126 square inches. A 13 × 13-inch flue lining, outside measurement, has a gross inside area of 127 square inches. The effective area would be only 100 square inches, which is somewhat smaller than should be used and may cause the fireplace to smoke.

In new chimney construction, the area of the flue can be calculated and built to satisfy the need of the fireplace opening. On existing chimneys, increasing the cross sectional area of the flue means tearing down the entire chimney and building a new one—highly impractical in most cases. The more practical solution is to decrease the size of the firebox. Install a metal hood across the top of the fireplace opening. A hood constructed of sheet copper can be quite attractive. If you don't have the tools to build it yourself, your local sheet metal shop should be able to handle the job for you at a moderate cost.

The easiest way to reduce the size of the fireplace opening with masonry is to build up the floor of the fireplace with one or two layers of bricks. This is how I corrected a smoky fireplace in my last home. After making the necessary calculations, I determined that one layer of bricks placed in the bottom of the fireplace will reduce the fireplace opening enough to allow for proper draft. I extended the bricks outside the fireplace and onto the hearth. This resulted in a very attractive raised hearth in front of the fireplace opening.

If the extra layer or two of bricks does not cure the problem (and everything else is okay), then line the sides and the back wall with firebricks. By trial and error you will eventually come up with the correct size fireplace opening that will allow your chimney to draw as it should. Be certain that your flue is undersized for your

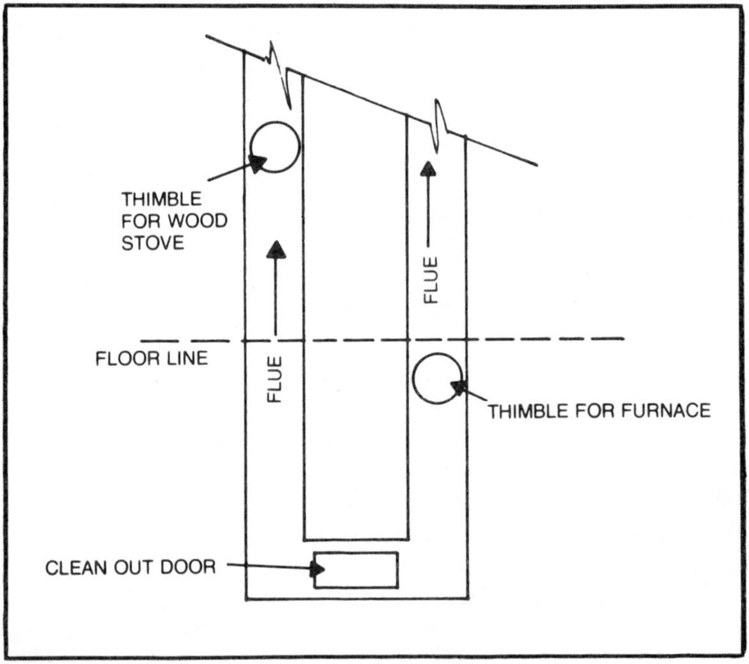

Fig. 7-1. Two flues, one for the oil-fired furnace and one for a wood stove on the next floor, are joined together at the bottom. Draft problems will occur, but these can be corrected if the flues are sealed off from each other.

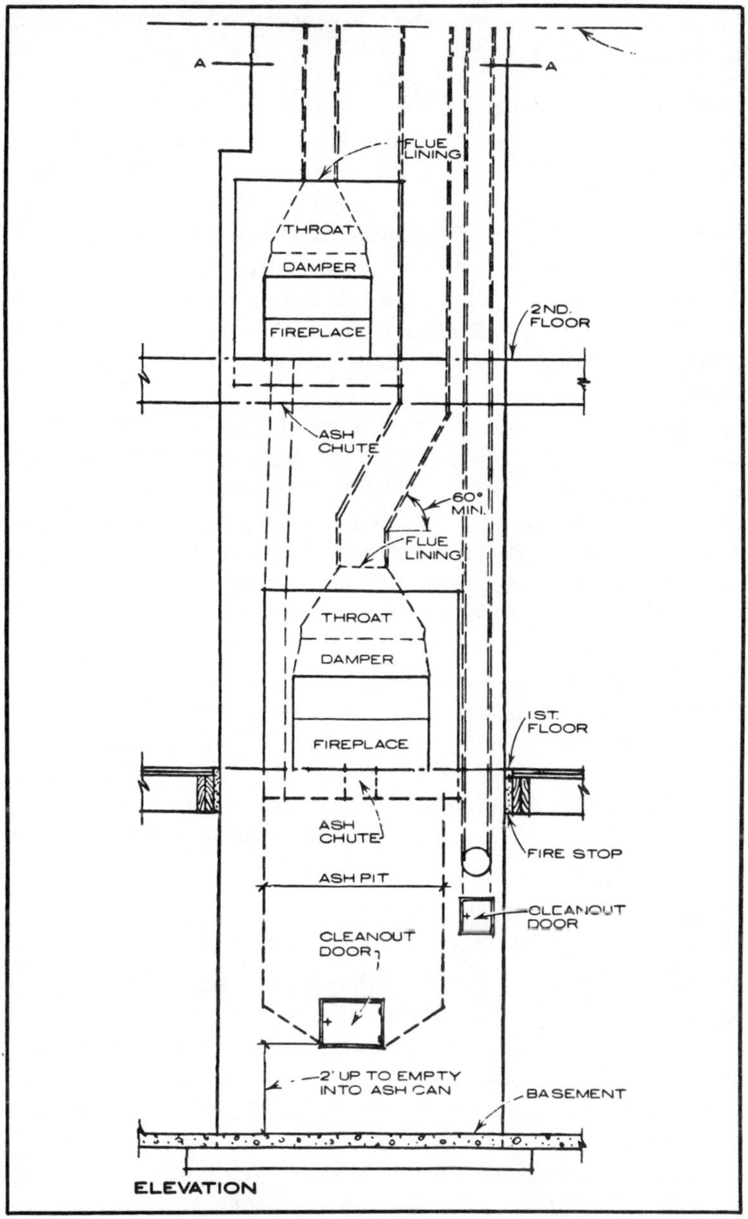

Fig. 7-2. Homes with multiple fireplaces and flues often contain only one cleanout door at one common location. Such arrangements can cause smoke problems and should be corrected by providing a cleanout door for each flue, and the opening between the flues should be sealed so that each flue is independent of the other.

fireplace opening; there may be other reasons why your fireplace or wood stove is smoky.

OTHER OPENINGS IN THE FLUE

In looking over your chimney for possible causes of a smoky fireplace or woodburning stove, see that there are no other openings into the flue, either above or below the woodburning stove. Check to see that the division walls of the chimney (if it contains more than one flue) are carried up to the top of the chimney and down to the bottom of the chimney, so that each flue is independent of the other. Air may also leak from one flue to the other through poorly sealed joints. The cure is to seal the joints as described in Chapter 2.

Another common problem is shown in Fig. 7-1. Here two flues are joined together at the bottom. The smoke pipe from the wood stove attached to the flue on the left will obtain air from the chimney on the right causing draft problems. It should be sealed off just below the opening where the stovepipe enters the flue.

Homes with multiple fireplaces and flues often contain only one cleanout door—all flues joining at one common location, which is usually in the basement of the home. Such an arrangement can cause smoke problems and should be corrected by providing a cleanout door for each flue. The opening between the flues should be sealed so that each flue is independent of the other (Fig. 7-2).

If there are two or more flues in the same chimney, they should be at different heights to prevent the draft from one flowing down the other (Fig. 7-3). A better solution is to install a hood and separate the two flues with a wythe (Fig. 7-4).

CHIMNEY CONSTRUCTION

Downdrafts are frequent problems with chimneys. To minimize downdrafts, the top of the chimney should be at least 3 feet above flat roofs and 2 feet above a roof ridge or raised part of the roof within 10 feet of the chimney. Under certain wind conditions, tall buildings or trees near the chimney top may still create downdrafts. Increasing the height of the chimney, or adding a booster fan or turbine at the top, may be a solution.

A hood will prevent gusts of wind from blowing down the chimney. When hoods are installed, you must make certain that the total area left open must be at least equal to the cross sectional area of the flue. There are several ways to install a hood. One is to simply embed iron rods or pipes in each corner of the chimney top. The rods should be of a size to adequately hold a slab top and be

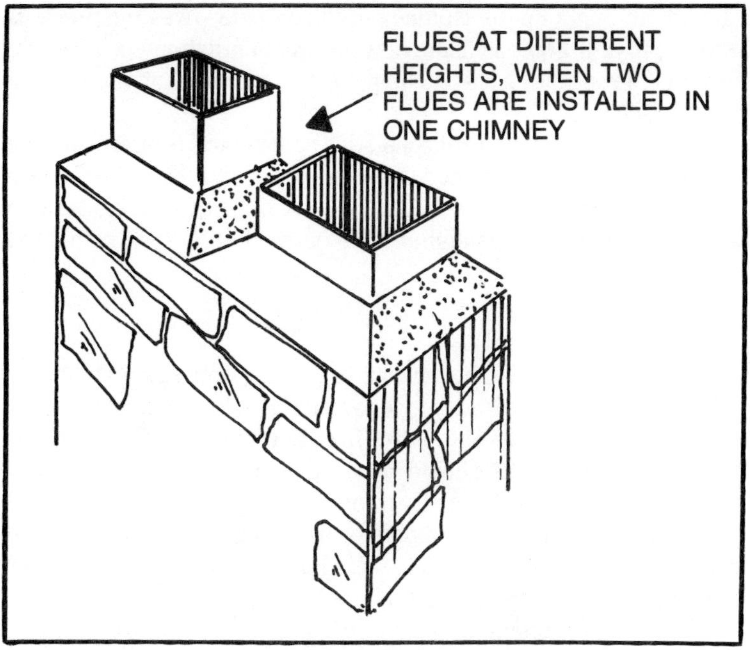

Fig. 7-3. If there are two or more flues in the same chimney, they should be at different heights to prevent the draft from one flowing down the other.

embedded into the masonry at the top of the chimney. A slate or cement slab may then be secured to the top of the rods.

The support may also be of masonry. When two flues are installed in the same chimney, a wythe should be provided between the two flues to prevent one flue from affecting the other. A hood placed above the chimney flue will also cut down on the possibility of sparks escaping.

Rather than going to the trouble or expense of changing the chimney's construction, many homeowners prefer to use some mechanical means to eliminate smoky stoves and fireplaces. One method is to install a booster fan or turbine at the top of the flue (Fig. 7-5). It is connected to a switch at a convenient location inside your home. When downdrafts or smoky fireplaces are nuisances, the fan is switched on and creates a suitable draft to prevent smoking of the fireplace or stove. In selecting such a unit, be sure to size it correctly. The supplier of the fan should be able to help. If too much draft is created, it will suck all the heat out of the living area faster than the fireplace can produce it. If accidently turned on during warm weather, it will suck all the cool air conditioned by your air conditioner right up the chimney to the outside—wasting precious energy. If your home is using only natural ventilation during warm weather, an operating fan will help create a cooling draft if you crack your windows.

A ceramic chimney pot consisting of two parts is offered by one manufacturer and is designed to increase updrafts. The inner cone is cemented to the top of the existing flue, and the chimney is built up around it a few courses higher. The outer shell is then cemented over it.

THE DAMPER

Your fireplace damper may have been improperly installed by using the front edge of the damper as a lintel to help support the row of bricks immediately above the fireplace opening. For a proper installation, the lintel should be a separate piece of angle iron located 6 to 8 inches lower than the damper. You can tear out the brickwork and raise the damper to the proper height, but a more practical solution would be to install a hood on the fireplace to lower the opening in relation to the damper.

On new construction, be sure to install the damper 6 to 8 inches above the top of the fireplace opening. Make it the same length as the width of the fireplace opening. If a shorter one has already been

Fig. 7-4. A hood will prevent gusts of wind from blowing down the chimney, but they must have an area equal to the cross sectional area of the flue. A hood above a chimney will also cut down on the possibility of sparks escaping and possibly causing forest fires.

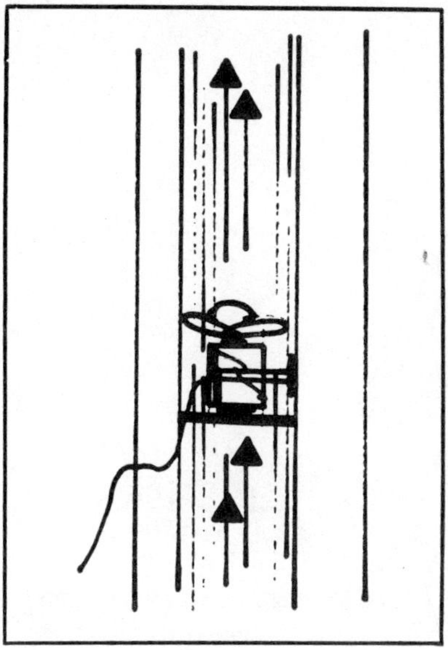

Fig. 7-5. When all else fails, a small fan can be installed in the flue to draw the smokes and gases through the flue. All electrical wiring must be approved for such use, and the insulation should be approved for the temperatures expected in the flue.

installed, you can build in the walls of the fireplace opening, making it narrower to match the damper size.

Some older homes do not have dampers in their fireplaces, and fire and draft control is difficult. A patented damper can be installed at the top of the chimney and controlled at the fireplace by a long rod running down the flue with a handle at the bottom. Smoking problems can be controlled with this device.

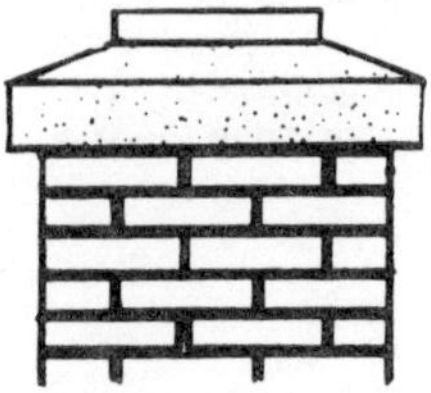

Chapter 8

Codes and Standards

There are certain state and local codes that dictate how a chimney must be built. While these differ from area to area, the following general description should suffice if you are contemplating building a new chimney in a new or existing home. The recommendations of the NFPA (National Fire Protection Association) also have developed certain standards for safe chimney building. You may purchase their bulletins by writing to NFPA, 470 Atlantic Ave., Boston, MA 02210.

The following suggestions do not apply to chimneys for use with high pressure boilers, furnaces used in manufacturing, or for other heating appliances where high temperatures are maintained. They apply to all other chimneys.

CHIMNEY CONSTRUCTION

The walls of all chimneys—regardless of the type of fuel to be used—should be built of brick, concrete, stone, or hollow tile, of such thickness and construction as is hereafter recommended. This does not preclude the use of a metal smokestack when a chimney is located inside a vent shaft having walls not less than 8 inches thick, and having an air space between the walls and stack on all sides.

Brick Chimneys

Brick chimneys should be built of solid bricks. They may be built of perforated radial brick manufactured for the purpose and adapted to withstand high temperatures, but no other hollow bricks should be used.

The walls of brick chimneys should be not less than 3-¾ inches thick (width of a standard size brick). They must be lined with fireclay flue lining.

Flue lining may be omitted in brick chimneys for private dwellings, provided the walls of the chimneys are not less than 8 inches thick. The inner course must be of a refractory clay brick having a softening point of at least 1922° F.

Perforated radial brick chimneys may be unlined, provided the brick has a softening point of not less than 1994° F. They should be not less than 7½ inches in radial thickness, except that when a chimney is located inside a vent shaft having walls not less than 8 inches thick, the thickness of the chimney wall may be determined by engineering design. The brick should be shaped to the circular and radial lines of the various sections of the shaft to form even joints.

All brickwork should be laid in spread mortar, with all joints push-filled. Exposed joints, both inside and outside, shall be smooth. No plaster lining is permitted.

Concrete and Stone Chimneys

Concrete chimneys cast in place shall be suitably reinforced vertically and horizontally. The walls should be not less than 3-¾ inches thick and be lined with fireclay flue lining.

Flue linings may be omitted in reinforced concrete chimneys for private dwellings, provided the walls of the chimneys are not less than 6 inches thick, quartz gravel is not used as the coarse aggregate.

Concrete blocks used in chimney construction should have walls not less than 3-¾ inches thick. Blocks enclosing more than one flue should have suitable reinforcement completely encircling the blocks and well embedded in them. All concrete block chimneys should have fireclay flue lining.

Stone chimneys should be at least 4 inches thicker than required for corresponding brick chimneys. They should have fireclay flue linings. Rubble stone chimney walls should be not less than 12 inches thick.

Walls

Hollow building tile should not be used for the walls of isolated or independent chimneys. It may be used for chimneys built in connection with exterior hollow tile walls of buildings not exceeding three stories in height, in which case the chimney walls should

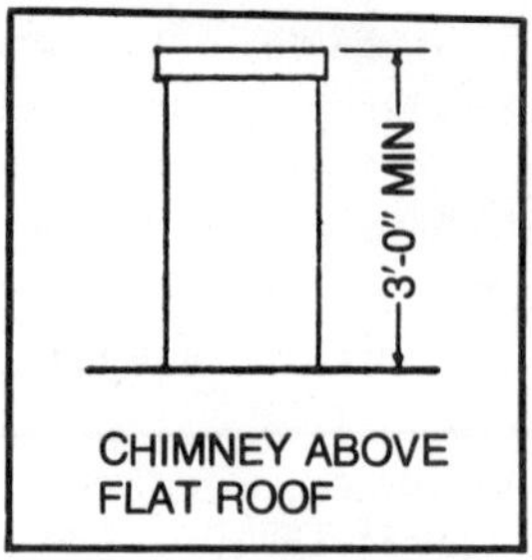

Fig. 8-1. Chimneys projecting above flat roofs should be at least 3 feet above the roof.

be not less than 8 inches thick. The outer 8 inches of a building wall may serve as the outside wall of the chimney, but the remaining chimney walls must be constructed of two layers of 4-inch tile set with broken joints. They may be built of 4 inches of solid brickwork. In either case, the walls of the chimney must be securely bonded into the wall of the building. No chimney should be corbeled from a hollow wall of tile. All chimneys or hollow building tile should have fireclay flue lining.

Chimneys should be built at least 3 feet above flat roofs and 2 feet above the ridges of peak roofs. They should be properly capped with stone, terra-cotta, concrete, cast iron, or other approved material. No cap should decrease the required flue area (Figs. 8-1 and 8-2).

Fireclay flue linings should be manufactured from suitable refractory clay, either natural or compounded, which has a softening point not lower than 1994° F. The linings should be adapted to withstand high temperatures and flue gases. They should be of standard commercial brick thickness, but not less than ¾ inch. The flue sections should be set in quality mortar and have the joints struck smooth on the inside. The masonry should be built around each section of lining as it is placed, and all spaces between masonry and linings shall be completely filled with mortar. No broken flue lining is permitted. Flue linings should start at least 4 inches below the bottom of the smoke pipe intakes of flues or from the throats of fireplaces. They should project at least 4 inches above the chimney top to allow for a 2-inch wash and a 2-inch lining projection. The wash or splay should be formed of a rich cement mortar. To improve the draft, the wash surface should be concave wherever practical. Instead of the wash, a special chimney cap or coping may be used. Wherever flue linings are specified, firebricks may be substituted if desired.

Chimneys should not rest upon or be carried by wooden floors, beams, or brackets; nor should they be hung from wooden rafters.

Iron brackets or stirrups attached to wood construction shall not be used to support chimneys. In frame buildings, chimneys must always be built from the ground up or rest on basement walls.

Chimneys should be built upon concrete or masonry foundations properly proportioned to carry the weight imposed without danger of settlement or cracking. The foundation for an exterior chimney shall start below the frost line.

The walls of brick buildings may form a part of a chimney, but the walls of the chimney must be securely bonded into the walls of the building. The flue should be lined the same as an independent chimney. Flues in party walls should not extend beyond the center of the walls, and their location should be permanently indicated on the exposed side of the wall.

No wall less than 12 inches thick should support a corbeled chimney. Such corbeling should not project more than 6 inches from the face of the wall. The corbeling should consist of the least five courses of bricks.

Flues

Flues should be built as nearly vertical as possible, but in no case shall they have an angle greater than 45° from the vertical (Fig. 8-3). Where flues change direction, the abutting linings at the angle joints must be chipped to fit closely. At no point should the cross section area be reduced. There should be but one connection to each flue.

Not more than two flues should be permitted in the same flue space, and the joints of any two adjoining sets of flue linings should be offset at least 7 inches. When there are more than two flues in a chimney, at least each third flue should be separated from the others by a wythe or division wall of brick or concrete at least 3-¾ inches thick and bonded into the sidewalls. Each flue intended for a heating

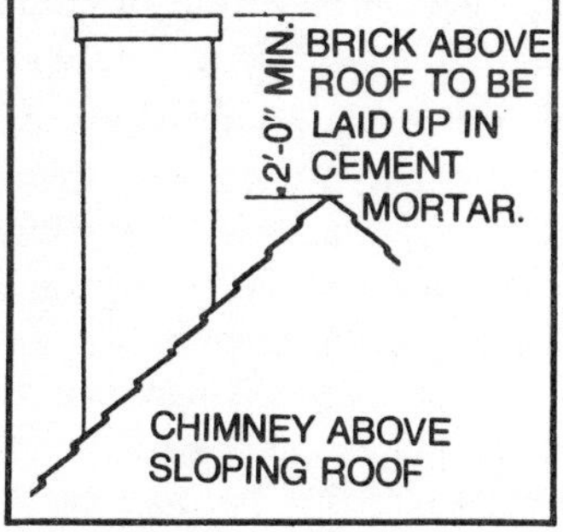

Fig. 8-2. Chimneys above the sloping roof should be a minimum of 2 feet above the roof's peak.

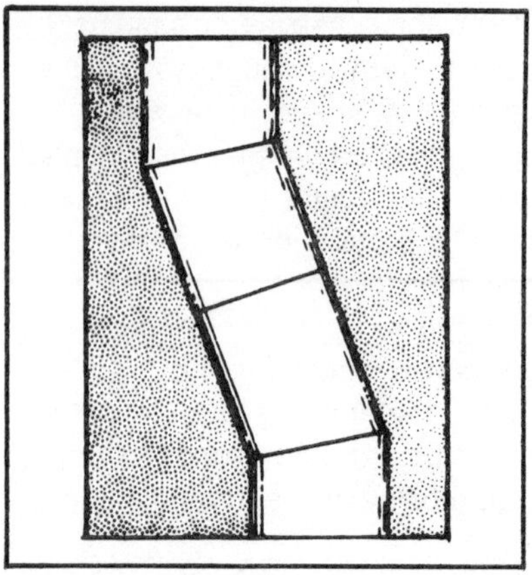

Fig. 8-3. Flues should be built as nearly vertical as possible. Where an offset is required, in no case should the offset have an angle greater than 45° from vertical.

furnace, boiler connection, or fireplace should be separated from other flues by such a wythe. The wythe may be of tile in hollow tile chimneys.

When any single flue area within the chimney walls exceeds 200 square inches, the walls should be built not less than 8 inches thick. They should have fireclay flue lining as previously mentioned. When flues become so large as to render it impractical to secure fireclay lining, they should be lined with firebricks for a distance of at least 25 feet from the point of intake. Firebricks should be laid in fireclay mortar.

Connections between chimneys and roofs should be made with sheet metal counter or cap flashing (copper is highly recommended). It is arranged to overlap roof flashing and allow for movement that may occur between chimneys and roofs (Fig. 8-4).

No increase in the wall thickness of chimneys, nor any projecting masonry, should be permitted within a distance of 12 inches above or below the rafters or roof joists.

Regardless of the type of fuel—coal, coke, wood, or oil—the minimum area inside of chimney flue linings for various heating devices should be as follows:

- For warm air furnaces, or low pressure steam or hot water heating boilers, not less than 75 square inches.
- For fireplaces, not less than one-tenth the area of the fireplace opening, but never less than 75 square inches.
- For stoves, ranges, and other forms of room heaters, 49 square inches for rectangular flues, or an inside diameter of 7 inches for round flues. In no case should the short cross section dimension of a rectangular flue be less than two-thirds the greater dimension.

When gas is the fuel used in a heating furnace, boiler, or automatic hot water heater, the flues should be of the same size and construction as required for stoves and ranges using other fuel. Vent flues where required for other domestic gas burning appliances may be of smaller size, but not less than 10 square inches. Such flues should be made of fireclay or its equivalent not less than 1 inch thick, with joints properly designed to effect a permanent seal. The surrounding masonry wall may be omitted. Metal vent flues are not permitted in most areas.

Smoke pipe intakes to flues should always enter the chimney through the side and consist of fireclay or metal thimbles securely set in the chimney wall with mortar, or the intake may be cast in concrete (Fig. 8-5). Such openings should be at least 18 inches

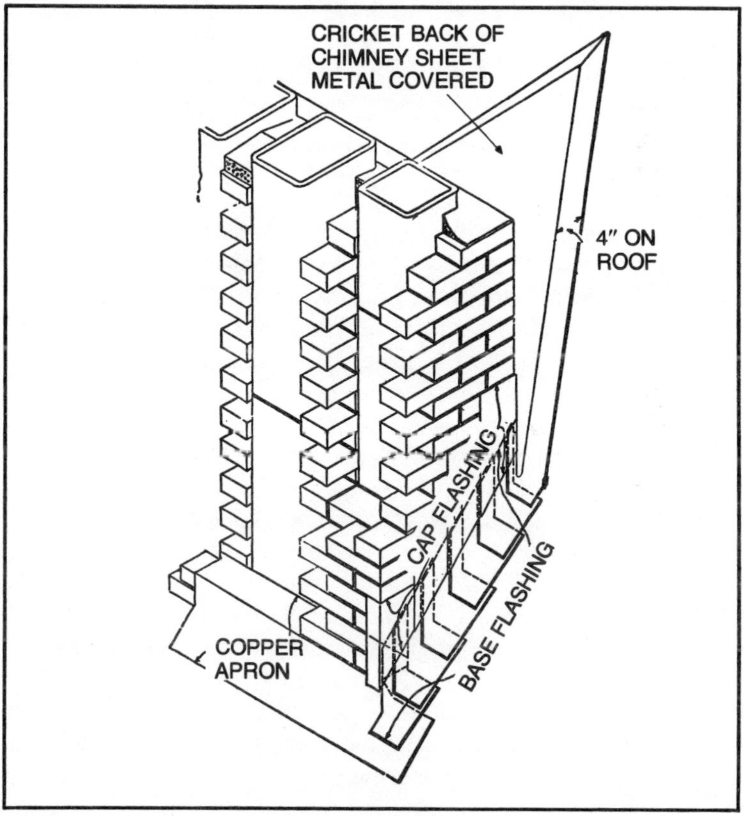

Fig. 8-4. Connections between chimneys and roofs should be made with sheet metal counter or cap flashing arranged to overlap roof flashing and allow for movement that may occur between the chimney and roof.

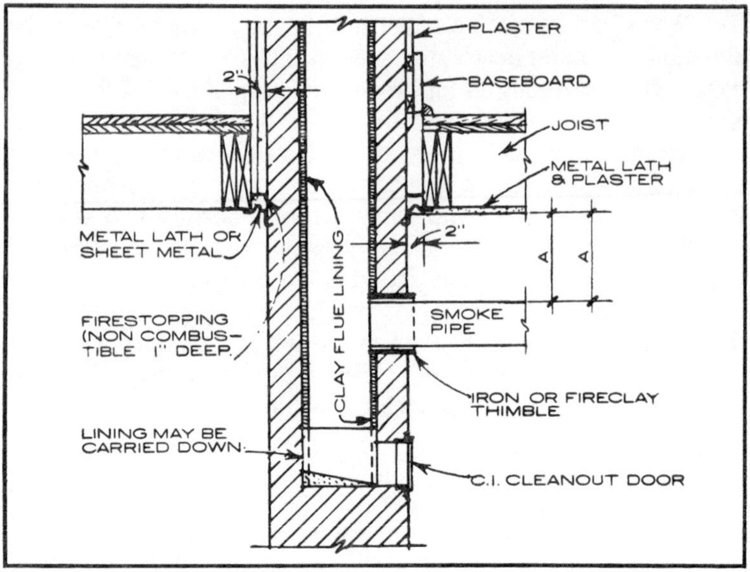

Fig. 8-5. Smoke pipe intakes to flues should always enter the chimney through the side and should consist of fire clay or metal thimbles securely set in the chimney wall with mortar.

below wooden lath and plaster or other combustible ceilings, or open joists. Neither the intake pipe nor thimble should project into the flue. No woodwork should be placed within 6 inches of the thimble. The thimble should be surrounded by metal lath and plaster for a space of at least 6 inches. An open space of that width should be provided on all sides.

Mortar

All mortar used in chimney construction, except as specified for firebricks, should be cement mortar proportioned as follows: two bags of portland cement, not less than 188 pounds, and one bag of dry hydrated lime, 50 pounds, thoroughly mixed dry. To this mixture should be added three times its volume of clean, sharp sand with sufficient water to produce proper consistency. When dry hydrated lime is not available, 1 cubic foot of completely slacked lime putty may be substituted for 50 pounds of dry hydrate. The mixing of lime and cement should be very thorough. Dry hydrate should always be used in preference to lime putty.

Smoke Test

After a chimney has been completed, all flues should be thoroughly cleaned and left smooth on the inside. All flues to which heating furnaces or boilers are to be connected should be subjected to a smoke test before use. The test should not be made until the mortar has thoroughly hardened. The test method is to build a smudge fire at the bottom of the flue. While the smoke is flowing freely from the flue, close it tightly at the top. Escape of smoke into other flues or through the chimney walls indicates openings should be made tight before the chimney is accepted. The contractor will make the test in the presence of the building inspector or other official having jurisdiction, and the owner or his representative.

WOODWORK AROUND CHIMNEYS

No wooden beams, joints, or rafters should be placed within 2 inches of the outside face of chimneys. No woodwork is placed within 4 inches of the back wall of any fireplace (Fig. 8-6).

All spaces between chimneys and wooden joists or beams should be filled with loose cinders, loose mortar refuse, gypsum block, or other porous incombustible material to form a fire stop (Fig. 8-7).

The incombustible material should be supported by strips of sheet metal or metal lath set into the brickwork and nailed to the wooden beams, forming a buckled flexible joint between, or by similar strips of metal nailed to the woodwork with the inner edge close to the chimney (Fig. 8-8).

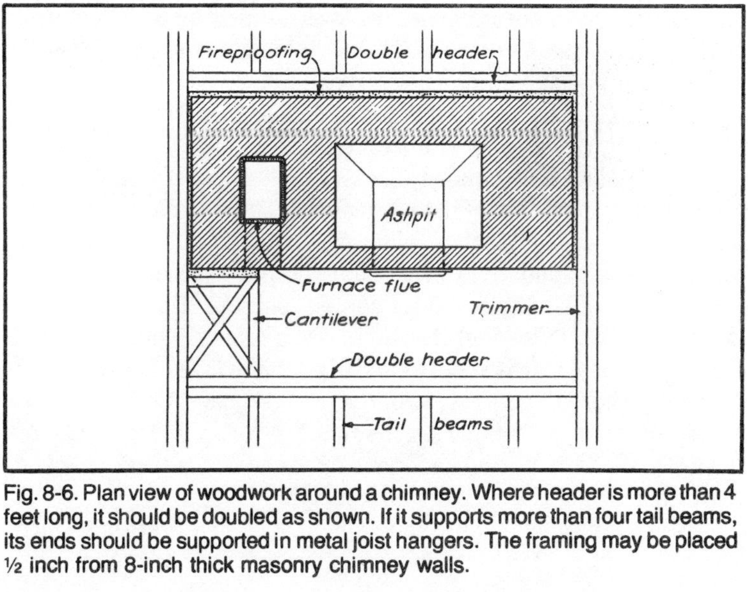

Fig. 8-6. Plan view of woodwork around a chimney. Where header is more than 4 feet long, it should be doubled as shown. If it supports more than four tail beams, its ends should be supported in metal joist hangers. The framing may be placed ½ inch from 8-inch thick masonry chimney walls.

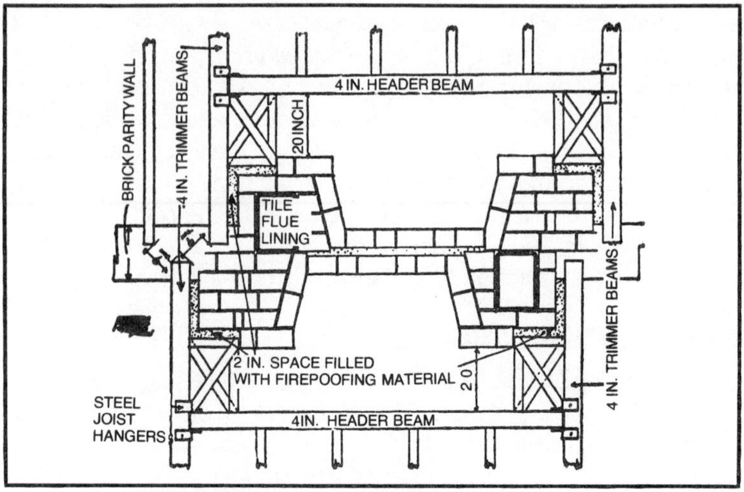

Fig. 8-7. Method for building two fireplaces back to back in a brick party wall to secure proper spacing between the ends of floor joists.

No wooden studding, furring, lathing, or plugging should be placed against any chimney or in the joints thereof. Wooden construction should either be set away from the chimneys, or the plastering should be directly on the masonry, on metal lathing, or on incombustible furring material. Wood furring strips placed around chimneys to support the base or other trim should be insulated from the masonry by asbestos paper, at least ⅛-inch thick. Metal wall plugs or approved incombustible nail holding devices attached to the wall surface should be used for nailing.

The walls of fireplaces will never be less than 8 inches thick. If they are built of stone, the minimum thickness will be 12 inches.

All fireplaces and chimney breasts should have trimmer arches or other approved fire-resistant hearths. The arches and hearths should be at least 20 inches wide, measured from the face of the chimney breast, and be of brick stone hollow tile not less than 4 inches thick. A flat stone or a reinforced concrete slab may be used to carry the hearth instead of an arch, if it is properly supported and a suitable fill is provided between it and the hearth. The length of trimmer arches and hearths should be not less than 24 inches longer than the fireplace opening. Hearths should be of brick, stone, tile, or concrete. Wood centering under a trimmer arch should be removed before plastering the ceilings underneath.

No coal burning heater should be placed in a fireplace which does not conform to the foregoing requirements and have an incombustible mantel. No wooden mantel or other woodwork should be placed within 8 inches of the side or 12 inches of the top of any open fireplace. Note the floor framing around a single fireplace in Fig. 8-9.

SPECIFICATIONS

Written specifications furnished with working drawings by engineers and architects usually accompany a set of working drawings for a new building or the renovation of an existing building. Heating specifications normally come under Division 15—Mechanical, but work for chimneys and fireplaces will normally fall

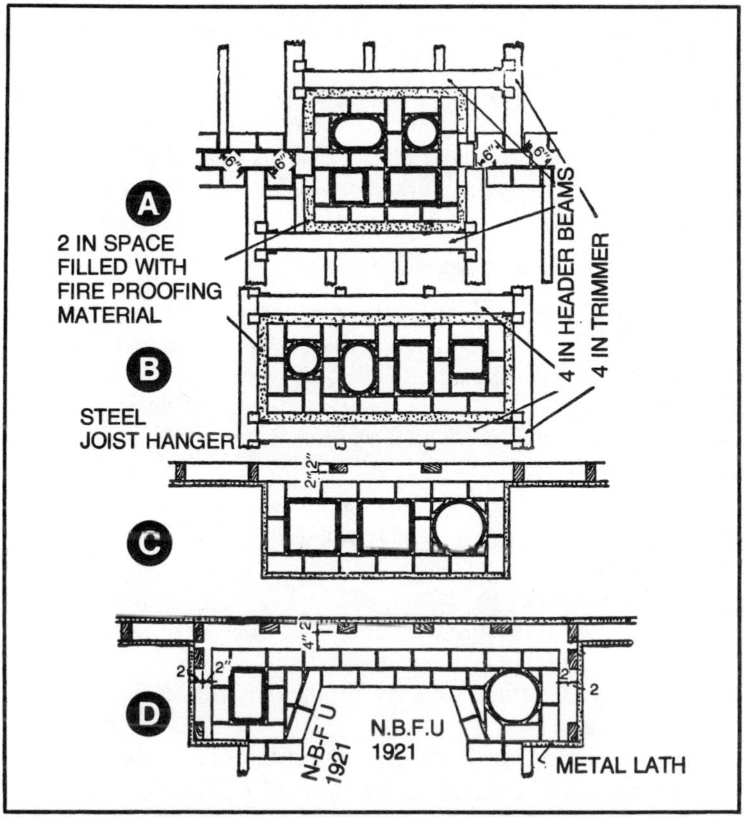

Fig. 8-8. (A) Floor framing around a chimney in a party wall to secure proper space between ends of floor joists. (B) Ordinary floor framing around a chimney. All timbers 2 inches clear of brickwork and space are filled with fireproofing material. (C) Stud partition across the back of a chimney showing the proper method of arranging studs. (D) Stud partition across the back of a fireplace and around the ends of the chimney breast, showing proper arrangement of studs.

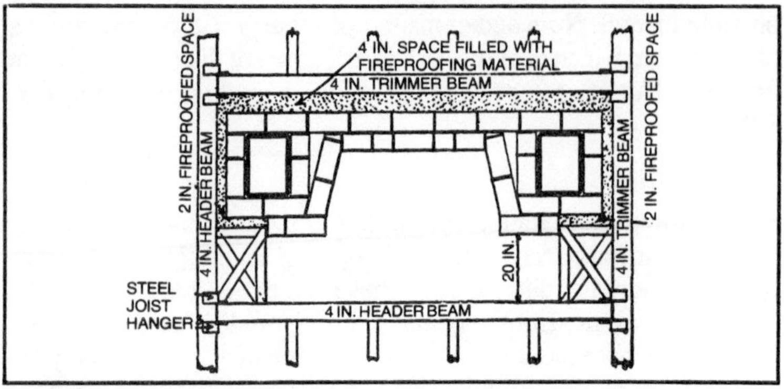

Fig. 8-9. Floor framing around a single fireplace. Note filling between framing and brickwork, which serves both as an insulator and fire stop.

under *Division 4–Unit Masonry.* To give an example of an architect's written specifications, I have furnished the following material taken from an actual project. You may wish to follow these specifications in doing the work yourself, or use them as guidelines to check a contractor who may be doing the work for you.

SECTION 4B: UNIT MASONRY

1. Portions of the sections of the documents designated by the letters "A," "B," and "C" and "Division One—General Requirements" apply to this division.

2. Consult the index to be certain that the set of documents and specifications is complete. Report omissions or discrepancies to the architect.

3. Standards must meet requirements and recommendations of applicable portions of the standards listed.

    (1) American Society for Testing and Materials (ASTM).
    (2) National Concrete Masonry Association (NCMA).
    (3) Structural Clay Products Institute (SCPI).

4. **Protecting**

a. Provide for heating and protecting masonry work in cold weather. Do not use frozen or ice-covered materials.

b. Do not lay bricks or blocks when the temperature is below 40° F. except with specific approval. Keep masonry work moist, in hot weather, until the mortar has cured.

c. Cover tops of walls at the end of the day's work. Protect finished masonry work from damage.

d. Protect persons from injury and property from damage. Satisfactorily repair or remove and replace work that has been damaged. Remove and replace discolored masonry work that will remain exposed.

5. Delivering and Storing

a. Arrange deliveries to provide sufficient quantities for continuity of any phase of work. Do not change suppliers or brands without specific approval.

b. Store bricks to prevent damage to materials or structure. Store bricks and blocks under a cover that permits circulation of air.

6. Materials

a. Common brick shall be first quality standard size hard burned common brick, uniform in size and straight. Sample is to be submitted for approval.

b. Face brick shall be of approved manufacturer, color and texture, standard sized brick. This contractor shall allow in his bid, sums as specified in Division 1, Section 1H, per thousand for face brick delivered on the job site.

c. Concrete block shall be standard hollow or solid concrete units (*Solite* or equal) of approved manufacture, and shall be of proper sizes to bond. Units shall conform to the following standards:

(1) Hollow load-bearing concrete masonry units—ASTM C 90.

(2) Hollow non-load-bearing concrete masonry units—ASTM C 129.

(3) Solid load-bearing concrete masonry units—ASTM C 145.

(4) Concrete building brick—ASTM C 55.

(5) Reinforced masonry shall be high pressure steam-cured, expanded aggregate lintels of designs shown on the structural drawings.

Note: hollow concrete masonry units filled with mortar will not be accepted as solid units. Concrete masonry units shall have a minimum moisture content, at the time of delivery to site, of 30 percent (of the total absorption).

d. Reinforcing

(1) All bearing or exterior masonry walls shall be reinforced with horizontal wall reinforcing at 16 inches on center vertically above exterior finished grade, and 8 inches on center vertically below finished grade.

(2) Corrugated ties are to be ⅞ inches wide, 6 inches

long, 16 gauge zinc-coated sheet steel. Cavity ties are required where noted.

(3) Horizontal wall reinforcing is to be of zinc-coated cold drawn steel wire, with the sides of deformed 9 gauge and smooth 12 gauge wires butt welded to the sides in same plane. Size for wall thicknesses shown.

(4) Column wall ties shall be zinc-coated cold drawn 3/16 inch steel wire.

(5) Concrete anchors shall be 20 gauge galvanized dovetail with corrugated ties. Cavity ties are required where noted.

7. **Sample panels**

a. Lay up, where directed at the site, sample panels not less than 2 feet high by 3 feet wide, using the full range of bricks proposed for the facing.

b. Do not proceed with brickwork until a sample panel has been approved. Demolish panels that are not approved.

c. Use an approved panel as the standard of comparison for workmanship and materials.

d. Do not destroy, alter, or move the panel until the brickwork has been approved as completed.

e. On completion of the project, remove the panel from the site and clean the space that it occupied.

8. **Workmanship**

a. Lay blocks and bricks plumb and true to a line, in a running bond or as indicated. Align on the exposed face or as indicated. Exposed cells, cracked blocks, and broken blocks are not permitted in work that will remain exposed.

b. Finish flush block joints that will not remain exposed. Rub joints with burlap, where blockwork will have additional finish, to make the joints match the texture of the block. Use concave tooled joints for masonry work that will remain exposed. Tool vertical joints first. Joints shall be of a width to suit the coursing with vertical joints of the same size as horizontal.

c. Dampen blocks and concrete bricks on mortar faces only, in hot weather, immediately before laying. Dampen brick, in hot weather, immediately before laying. Lay masonry without units carrying surface water.

d. Lay hollow blocks in a full mortar bed on shell surface and at ends. Lay concrete bricks and solid blocks in a full mortar bed with head and edge joints completely filled. Lay bricks in a full bed of mortar with head joints and edge joints completely filled.

e. Cut units with a power saw where block will remain exposed. Lay out courses to minimize cutting and to avoid jumping bond.

f. Cut bricks with a power saw where necessary for fitting or bonding. Broken units are not permitted in exposed faces. Lay out courses to minimize cutting and to avoid jumping bond.

g. Bond intersecting and abutting walls with interlocking alternate masonry units or with corrugated metal anchors at 16 inches on center. Where a non-bearing wall abuts a bearing wall, provide corrugated metal anchors at 8 inches on center. Do not bond the two walls together. Leave a minimum ¼-inch joint and pack it with oakum and caulk.

9. **Cutting and repairing**

a. Do not cut installed materials and items without prior approval.

b. Cutting, when approved, and repairing shall be done by the trade that installed the original work and to avoid damage to adjacent work.

c. Repair work that will remain exposed in a manner that will not be obvious as a patch. Remove and replace work that cannot be satisfactorily repaired.

10. **Pointing and cleaning**

a. General: on completion, all exposed surfaces shall be cleaned by mechanics skilled in such work. Any material liable to damage shall be protected.

b. Brickwork: Defective exposed joints shall be cut to a depth of ¾ inch and properly repointed. Mortar drippings and other foreign matter shall be removed from walls by a stiff brush immediately following the setting of the bricks and blocks. The surface shall be thoroughly wetted prior to cleaning with clean water. The surface shall then be washed with a solution of 1 part *Sure-Klean Masonry Cleaning* solution as manufactured by the *Process Solvent Company, Inc.,* Kansas City, KS, and 6 to 8 parts clean water, brushed with short fibered masonry wall washing brushes. Wash the wall thoroughly with clean water. A minimum drying out time of 28 days shall elapse between finishing the wall and cleaning. Windows, doors, and trim areas and planting shall be protected as directed by the architect.

11. **Related work**

a. Built-in work: work required to be built into masonry walls shall be installed as the work progresses, and shall be solidly grouted under and filled around. This applies to sills, frames, anchors, etc., which are to be furnished by other trades.

b. Chases: The contractor shall build in all chases, frames, sleeves, bolts, etc., as required on the job whether these items are furnished by him or by others.

DIVISION 3: CONCRETE, SECTION 3A: FORMWORK

Other sections of architects' specifications that apply to chimneys and fireplaces include mortar and concrete. Examples of these sections follow.

1. Portions of the sections of the documents designated by the letters "A," "B," and "C" and "Division One—General Requirements" apply to this division.

2. Consult the index to be certain that the set of documents and specifications is complete. Report omissions or discrepancies to the architect.

3. Standards meet requirements and recommendations of applicable portions of the standards listed.

(1) American Concrete Institute (ACI).

4. **Complying**

a. Comply with applicable requirements of ACI 301-66 "Specifications for Structural Concrete for Buildings."

b. Comply with applicable recommendations of ACI 347 "Recommended Practice for Concrete Formwork."

5. **Materials**

a. Exposed concrete surfaces: ⅝-inch or ¾-inch thick, exterior grade, medium or high density resin fiber overlaid Douglas fir plywood.

b. Concealed concrete surfaces

(1) Plywood: DFPA-EXT "Plyform", Grade BB, 5.8-inch or ¾-inch thick, as manufactured for formwork.

(2) Boards: "T&G," commercial-standard Douglas fir or number 2 common or better lumber, uniform, nominal 1-inch thickness to provide tight forms.

c. Use adjustable form ties with working strength of not less than 3000 psi. Metal is not permitted closer than ¾ inch to finished surfaces. Do not use ties or spreaders that will leave a hole larger than ⅞ inch in exposed surfaces. Wire ties are not permitted.

6. **Construction**

a. Construct forms to shapes, lines, and dimensions shown. Securely brace forms to prevent displacement and to safely support construction loads. Forms shall be substantial and tight to prevent leakage. They shall be braced or tied to maintain their position and shape where concrete is tamped or vibrated. Forms for all exposed parts of concrete shall produce a smooth surface on the concrete with formed chamfer or fillet corners.

b. Forms shall be erected to the exact lines and elevations and be carefully checked immediately prior to the placing of concrete. All form lumber shall be thoroughly cleaned before reusing, and lumber that is warped, cracked, or otherwise damaged shall not be reused. Patching of forms will not be permitted on exposed surfaces. All openings, offsets, risers, brackets, haunches, depressions, curbs, and the like shown or required in the concrete work shall be provided in the formwork. Anchor bolts shown or specified to be set in concrete shall be accurately located to templates. Build into forms all inserts, anchors, chases, sleeves, etc., as may be required for other work.

c. Forms shall be cleaned of all debris before concrete is placed. Allow ample time to place anchor bolts, sleeves, inserts, etc., in the formwork.

d. All shoring shall be of solid wood timbers of single lengths wedged into place, or patented steel pipe shores with a screw jack adjustment. Adjustable wood shoring will not be permitted.

e. Where soil conditions permit excavation to accurate sizes without cave-ins, side forms for footings may be omitted upon the architect's approval.

f. Where drawings indicate concrete work over gravel or earth, omit wooden forms and shape gravel or earth to form the concrete accurately as detailed.

7. **Wetting and oiling forms**

Inside surfaces of wood board forms shall be soaked with clean water prior to placing concrete. Plywood forms shall be treated with form oil so that forms are oily to the touch. Do not vary from the type of form oil used throughout the project.

8. **Removal of forms**

a. Forms shall be removed in such a manner as to insure complete safety of the structure. Care shall be taken to avoid spalling concrete surfaces.

b. Sides or face forms for walls, beams, columns, piers, and similar vertical surfaces may be removed after 24 hours, provided the concrete is sufficiently hard not to be injured thereby. Sufficient hardness shall be determined by a laboratory test of field cured cylinders. Forms for supported slabs shall remain in place at least until the end of the curing period.

SECTION 3B: EXPANSION AND CONTRACTION JOINTS

1. Portions of the sections of the documents designated by the

letters "A," "B," and "C" and "Division One—General Requirements" apply to this division.
2. Consult the index to be certain that the set of documents and specifications is complete. Report omissions or discrepancies to the architect.
3. Standards meet requirements and recommendations of applicable portions of the standards listed.
(1) American Concrete Institute (ACI).
(2) American Society for Testing and Materials (ASTM).
4. **Materials**
a. Expansion joint: ASTM D 1751-65, ½-inch preformed expansion joint filler.
5. **Expansion joints:**
These joints shall be provided where shown on the drawings or called for in the specifications, between concrete paving and other structures and where slabs abut a wall. Expansion joints between floor slabs and walls shall extend from the bottom of the slab to within ⅜ inch of the top of the slab in all interior spaces. The top ⅜ inch shall be filled with grout.

SECTION 3C: CONCRETE REINFORCEMENT

1. Portions of the sections of the documents designated by the letters "A," "B," and "C" and "Division One—General Requirements" apply to this division.
2. Consult the index to be certain that the set of documents and specifications is complete. Report omissions or discrepancies to the architect.
3. Standards must meet requirements and recommendations of applicable portions of standards listed.
(1) American Concrete Institute (ACI).
(2) Concrete Reinforcing Steel Institute (CRSI).
(3) American Society for Testing and Materials (ASTM).
4. **Complying**
a. Comply with applicable requirements of ACI 301-66 "Specifications for Structural Concrete for Buildings."
b. Comply with applicable requirements of ACI 318-63 "Building Code Requirements for Reinforced Concrete."
c. Comply with CRSI 63 "Recommended Practice for Placing Reinforcing Bars."
5. **Shop drawings**
a. Submit five copies of reinforcing steel shop drawings for

approval prior to fabricating bars. Drawings shall show size, lengths, bends, and placing and support details.

6. **Materials**

a. Reinforcing bars: ASTM A615 Grade 60, free of rust, scale, or coatings which reduce bond—sized as noted on the drawings.

b. Welded wire fabric: ASTM A185 of sizes as noted on the drawings.

c. Metal form over steel joists: shall be "Steeltex" as manufactured by National Wire Products Corporation.

7. **Placing Reinforcement**

a. Reinforcing steel shall be free of oil, grease, loose scale, rust, or other bond inhibiting agents. It shall be *cleaned* of such material before placement of concrete around the steel.

b. All bars shall have minimum cover described in ACI 318-63, Paragraph 808, unless otherwise noted.

c. Care shall be taken to prevent puncture of the vapor barrier.

d. Place reinforcement accurately in the position shown. Securely fasten and support it to prevent displacement before or during pouring. Cleaning, bending, placing, and splicing of reinforcement shall be done according to the requirements of ACI Standard ACI 301-66 and approved shop drawings.

e. Welded wire fabric reinforcement in slabs shall have sides and ends lapped not less than one mesh plus 2 inches, and positioned in the center of the slab.

f. "Steeltex" shall be rolled across the joists from the bearing wall of the structure and fastened every 12 inches on the first three joists of the anchor end. It shall be stretched and fastened to the last three joists in the same manner with not less than two clips per width on each intermediate joist—alternating direction of adjacent clips. Ends shall be spliced only over a joist with at least a 12-inch lap and no consecutive splices over the same joist. Provide bracing at sidewalls and bearing walls in accordance with manufacturer's recommendations.

SECTION 3D: CAST-IN-PLACE CONCRETE

1. Portions of the sections of the documents designated by the letters "A," "B," and "C" and "Division One—General Requirements" apply to this division.
2. Consult the index to be certain that the set of documents and specifications is complete. Report omissions or discrepancies to the architect.
3. Standards meet requirements and recommendations of applicable portions of the standards listed.

(1) American Concrete Institute (ACI).

(2) Concrete Reinforcing Steel Institute (CRSI).

(3) American Society for Testing and Materials (ASTM).

4. **Complying**

a. Comply with applicable requirements of ACI 301-66 "Specifications for Structural Concrete for Buildings."

b. Comply with applicable requirements of ACI 318-63 "Building Code Requirements for Reinforced Concrete."

c. Comply with CRSI 63 "Recommended Practice for Placing Reinforcing Bars."

d. Comply with ACI 315-65 "Manual of Standard Practice for Detailing Reinforced Concrete Structures."

e. Comply with ACI 614-59 "Measuring, Mixing, and Placing Concrete."

5. **Materials**

a. Portland cement: ASTM C150 Type 1, or Type 2.

b. Air entraining portland cement: ASTM C175-61 Type 1A. Use in exterior slabs and structures only.

c. Coarse aggregates: except as otherwise specified, aggregates shall be hard, durable, uncoated crushed stone, or gravel conforming to ASTM C33.

(1) Maximum size aggregate allowed is one-fifth of narrowest dimension between forms of the concrete member, or three-fourths of minimum clear spacing between reinforcing bars, or as recommended in ACI 613.

(2) For concrete having an exterior surface exposed, 95 to 100 percent shall pass a 1-½-inch sieve, 35 to 70 percent shall pass a ¾-inch sieve,
10 to 30 percent shall pass a ⅜-inch sieve,
and not over 5 percent shall pass a number 4 sieve.
Percentages are by weight.

d. Fine aggregate: ASTM C33 washed, sharp, natural sand.

e. Water: Clean and free of organic matter, acid, alkali, or other injurious elements.

6. **Manufacture**

a. Concrete shall be transit mixed in accordance with all requirements of ASTM C94 and the ACI 614, except as otherwise specified.

b. Below 50° F, prepare concrete in accordance with ACI 604.

c. Above 80° F, prepare concrete in accordance with ACI 605.

7. **Concrete quality**

a. Strength: see drawings for strength requirements.

(1) The contractor shall be responsible for using concrete of the specified qualities. Design mixes and compressive strength tests supporting same shall be submitted in advance for approval.

(2) All concrete, unless otherwise specified, shall consist of portland cement, coarse aggregate, fine aggregate, and water. No antifreeze or set accelerating admixtures will be permitted. The mixture shall develop a minimum ultimate 28 day strength equal to that noted on the drawings, except that the design strength of laboratory cured cylinder shall be 15 percent greater than the strength noted. No mixture shall develop ultimate 28 day strength less than 3000 pounds per square inch unless otherwise noted.

b. Slump: the slump of the concrete shall be the minimum that is practicable. When vibrators are used to consolidate the concrete, the slump shall not exceed 4 inches; otherwise, the slump shall not exceed 6 inches.

c. Air content: all concrete exposed to the elements shall have between 4 and 6 percent air entrained.

8. **Handling and placing**

a. All forms and bottoms shall be clean and free of water and debris before the concrete is placed. Except as otherwise noted, all reinforcing steel, sleeves, and other items to be embedded are in their proper places and wired in position before concrete is started. Thoroughly dampen forms before placing the concrete, except in freezing weather.

b. Ready mix concrete elapsed time from the time of departure from the plant until concrete is placed in forms is not to exceed 45 minutes. Delivery tickets shall show time of departure from the plant, proportions of cement, aggregate and water, and any water added at the job.

c. Concrete shall be conveyed as nearly as possible to its final position before it is released into the forms in accordance with ACI 318 and ACI 614. It is worked along the forms, along and around the reinforcing, into corners of forms, and is spaded and tamped to prevent the formation of voids. Maximum free fall of concrete shall be 3 feet. Chuting will be permitted only where the concrete is deposited into a hopper before it is placed in forms.

d. The concrete shall be vibrated with internal mechanical

vibrators of an approved type during placing. Vibration shall be sufficient to work concrete around the reinforcing, but not long enough to cause separation or segregation of the aggregates. Amount of vibration shall be in accordance with ACI 614.

e. Particular care shall be exercised in placing concrete which will be exposed in the finished work. The surfaces of all exposed concrete shall be smooth, even, and free from form marks or honeycomb.

f. When placing fresh concrete on work previously done, the surface of the old work shall be thoroughly cleaned of all scum and *laitance.* The surface is roughened by picking if necessary. The cleaned surface is wetted, all excess water removed, and a neat cement grout is broomed on just prior to placing new concrete.

g. After removal of forms, any honeycombed concrete shall be cut out as directed by the architect. Openings are filled with new concrete as required.

h. Note: take special care to adequately chair slab mesh.

i. Weather conditions

(1) Protection: unless adequate protection is provided and the architect's approval is obtained, concrete shall not be placed during rain, sleet, or snow. Rain water shall not be allowed to increase the mixing water or damage the surface finish.

(2) Placing temperature: cold weather shall be in accordance with ACI 306. Hot weather shall be in accordance with ACI 305.

j. Concreting shall generally be carried on as a continuous operation until the placing of an individual section is completed. When construction joints are necessary, they shall be made and located with the architect's approval.

9. **Tests**

a. The contractor shall engage an independent testing laboratory, subject to the architect's approval, to test the materials to be used in the mix, design the concrete mix, and test the concrete placed at the job. The concrete mixes shall be designed in accordance with number 7.

b. The contractor shall cast not less than three cylinders, 6-inch diameter × 12 inches high, from each pour or each 100 cubic yards, whichever is smaller. Cylinders shall be marked with a sign indicating the location of the concrete that they represent and the date cast. One of the three cylinders shall be tested at seven days, and the remaining two shall be tested at 28 days.

c. Compression test specimens shall be made in accordance with ASTM Designation C31. Specimens remaining at the site shall be cured in accordance with ASTM Designation C31. Laboratory curing of specimens will be done in accordance with ASTM Designation C192. Specimens will be tested in accordance with ASTM Designation C39.

d. Slump tests shall be made in accordance with ASTM Designation C143. Make such slump tests from truckloads as the laboratory shall deem necessary to verify the slump of concrete being poured.

The contractor shall include in his price the cost of mix design and testing of the concrete.

10. **Filling-in**

a. Fill in all holes and openings left in the concrete for pipe chases, ducts, steel, piping, sleeves, and other work unless otherwise shown or directed. Such parts shall be filled with concrete after the work of other trades is in place. All filling shall be neat and flush with adjacent surfaces.

b. The contractor shall provide all other miscellaneous concrete filling shown or required to complete the work as indicated.

c. All surface projections shall be removed. Joints, holes, and cracks shall be cut out and neatly pointed with cement mortar, mixed with bonding admixture per manufacturer's instructions. Honeycomb and porous surfaces shall be cut back to solid concrete and then filled with a cement mortar of 1 part cement to 3 parts sand. All metal ties, including form spreaders, shall be cut back to a depth of not less than ½ inch. Holes are pointed up.

11. **Finishing and curing**

a. Concrete shall be protected adequately from injurious action by the sun, rain, flowing water, frost, and mechanical injury. It shall not be allowed to dry out from the time it is placed until expiration of the minimum curing days. The temperature of the air next to the concrete shall be maintained at not less than 40° F for the full curing period.

b. Floor slabs

(1) All interior floors, except depressed areas to receive further finish, shall be screeded, wood floated, and steel troweled. A trowel shall be used until final troweling produces a shine or gloss to the surface of finish as required to make a smooth, hard, and impervious surface free of defects. Where floor drains occur, slope the slab uniformly to drain from

the walls of the room. Except for slopes to drains, floor slabs shall be level to within ⅛ inch in 10 feet.

(2) All interior floor slabs not receiving finish (resilient tile, paint, etc.), except for those areas to receive ceramic tile, shall be finished with Colorcron (natural color) floor hardener as manufactured by Master Builders at the rate of 1/3 pound per square foot. Floors shall be cured with MB-429 curing material as manufactured by Master Builders. All materials shall be applied in accordance with the manufacturer's directions.

(3) All other interior floor slabs are cured with MB429 curing material as manufactured by Master Builders in accordance with the manufacturer's directions.

c. Exterior

(1) Finish: all exterior concrete paving, loading dock slabs, sidewalks, stairs, etc., in addition to screening, floating, and troweling, shall be broom finished with a black bristle push type broom.

(2) Curing: slabs shall be fog sprayed immediately after finishing and covered with white pigmented polyethylene sheets sealed on all edges. Polyethylene shall be maintained in place for five to seven days, and no curing compound shall be used.

d. Vertical surfaces

(1) Rub all exposed vertical concrete surfaces with carborundum brick. Wet thoroughly all surfaces being rubbed, and maintain wet until rubbing is completed. Leave surface uniformly smooth and wash clean.

(2) Vertical surfaces from which forms have been removed shall be kept continually wet for the balance of the curing period.

12. **Unit prices:** this contractor in submitting his proposal shall give a unit price per cubic yard for concrete including reinforcing steel and forms as specified.

13. **Related work:** contractors for other trades shall furnish all anchors or other items which must be built into the concrete to receive their work. They shall furnish all information regarding the size and location of chases and openings and shall set all shores, sleeves, etc., for the proper application of their work.

14. **Cleaning:** clean all exposed concrete surfaces, and all adjoining work stained by the leakage of concrete, to meet the approval of the architect.
15. **Supervision:** Master Builders shall make available at no cost, upon 72 hours notification, the services of a qualified, full-time field representative to make clear to the user the application of the product under prevailing job conditions.

DIVISION 4: MASONRY, SECTION 4A: MORTAR

1. Portions of the sections of the documents designated by the letters "A," "B," and "C" and Division One—General Requirements" apply to this division.
2. Consult the Index to be certain that the set of documents and specifications is complete. Report omissions or discrepancies to the architect.
3. **Materials**

a. Masonry cement shall conform to ASTM Specification C91, Type II.

b. Portland cement shall conform to ASTM Specification C150, Type I.

c. Lime: hydrated lime shall be Type S or Type N, conforming to ASTM Specification C207.

d. Sand: shall be clean, natural, light colored and conform to ASTM Specification C144.

e. Water: shall be clean and fit to drink.

4. **Types of mortar**

a. Mortar for masonry in contact with earth shall be mixed in the proportions of 1 part portland cement, ¼ part hydrated lime, and between 2-¼ and 3 parts sand by volume. At the option of the contractor, mortar may be mixed in the proportions of 1 part portland cement, 1 part masonry cement, and between 4-½ and 6 parts of sand by volume.

b. Mortar for all unit masonry not specified otherwise shall be mixed in the proportions of 1 part portland cement, ½ to 1-¼ parts hydrated lime, and 4-½ to 6 parts of sand by volume. At the option of the contractor, mortar may be mixed in the proportions of 1 part masonry cement and 2-¼ to 3 parts of sand by volume.

5. **Mixing:** all materials shall be mixed a minimum of 3 minutes and a maximum of 5 minutes after all materials are in the mixer, with the amount of water required to produce the desired workability.
6. **Retempering:** if the mortar begins to stiffen from evaporation or from absorption of a part of the mixing water, retemper the mortar

immediately by adding water. Remix the mortar. All mortar shall be used within 2-½ hours of the initial mixing. It shall not be used after it has begun to set.

7. Delivering and storing

a. Arrange deliveries to provide sufficient quantities for continuity of any phase of work. Do not change suppliers or brands without specific approval.

b. Store cementlike materials in dry, weathertight, ventilated spaces. Store aggregates to exclude foreign matter and to maintain gradation.

Chapter 9
A Primer on Firewood

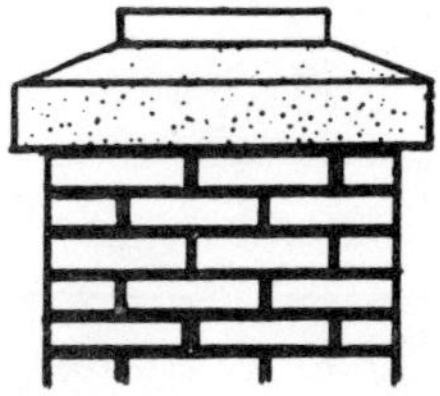

Firewood used for home heating should, if at all possible, be well-seasoned hardwood which has been dried at least 6 months. Seasoned wood produces less creosote and provides an average of 20 to 25 percent more usable heat (Table 9-1). Softwoods such as pines, spruces, cedars, and firs have a high resin content which can cause heavy creosote buildup. These woods should be avoided when possible except for kindling or perhaps an occasional aromatic fuel. Do not burn green wood, it also can create creosote problems. Green wood is easily spotted; it is heavier and looks different from the seasoned material. Seasoned wood usually has cracks radiating outward from the center toward the bark, while green wood normally has no crack pattern and will usually look newly cut.

HOW WOOD BURNS

The old technique of admitting air beneath the fire and letting it flow up under the fuel bed, and then directly into the chimney flue, was a major cause for the low efficiency and many problems commonly associated with woodburning. Wood burns quite differently from other fuels, and it needs specially designed combustion units to provide satisfactory heat.

Combustion of wood is preceded by evaporation of the moisture in the wood. Since it requires approximately 1,000 Btus (British thermal units) per pound of moisture to do this, it is important that the fuel be thoroughly seasoned. Then follows the distillation of volatile matter in the wood, beginning at about 400° F. After distillation is complete, wood is reduced to charcoal at about

Table 9-1. Evaluation of Some Common Woods Found in the United States.

species, listed in approximate order of density	available heat per cord green	available heat per cord air-dry	percent more heat from air-dry	price/cord[1] equivalent to heating oil at 50¢/gal average* fireplace	price/cord[1] equivalent to heating oil at 50¢/gal airtight* stove	price/cord equivalent to electric baseboard heat @4 ¢/KWH average* fireplace	price/cord equivalent to electric baseboard heat @4 ¢/KWH airtight* stove
	(millions of btu)			$	$	$	$
Hickory	20.7	24.6	19	20	68	43	144
White Oak	19.2	22.7	18	19	62	40	133
Sugar Maple	18.4	21.3	16	18	59	37	125
American Beech	17.3	21.8	26	18	60	38	128
Red Oak	17.9	21.3	19	18	59	37	125
Yellow Birch	17.3	21.3	23	18	59	37	125
Yellow Pine	14.2	20.5	44	17	56	36	120
White Ash	16.5	20.0	21	16	55	35	117
American Elm	14.3	17.2	20	14	47	30	101
Red Maple	15.0	18.6	24	15	51	33	109
Douglas Fir	13.0	18.0	38	15	49	32	105
Eastern White Pine	12.1	13.3	10	11	37	23	78
Aspen	10.3	12.5	21	10	34	22	73

[1] A full cord is a stack of wood 4 ft. high by 4 ft. wide by 8 ft. deep

* Assumed efficiencies: fireplace, 15%; airtight stove, 50%; oil furnace, 65%; electric baseboard, 100%

1,000° F. At this point the charcoal itself burns, leaving very little ash residue.

If efficient combustion is to occur, the volatile gases released during the process need to be mixed with air and burned completely inside the heating unit. Combustion of volatiles can account for as much as 60 percent of the heat energy in a log if the stove is designed to burn these gases. Unfortunately, the unburned volatiles frequently become a dangerous nuisance, as will be explained later. Successful burning of volatiles requires oxygen, a long flame path, and a temperature of about 1,100° F. Most old woodburning units do not make provisions for this, but allow gases to escape up the chimney where their heating value is lost. Better designs are needed before higher efficiency can be obtained.

If all of the air for combustion is admitted at the base and under the burning wood (primary air), most of the oxygen reacts in the bed of glowing charcoal to produce carbon monoxide, carbon dioxide, and heat. The heat rising through wood above the charcoal bed promotes release of volatiles, but these cannot burn since most oxygen in the primary air has been depleted in passing through the bed of coals. The unburned volatiles pass up the chimney, sometimes dissolving in condensed steam on chimney surfaces to form *pyroligneous acid* and, eventually, a gummy deposit called creosote. Acids cause corrosion of iron and steel, while deposits of creosote may ignite and fuel dangerous chimney fires.

Efficiency improvements can be obtained by designing secondary air inlets so that oxygen can be introduced and mixed with unburned volatiles without cooling them below their ignition point. Complete combustion can then occur. Industrial wood-fired boilers are designed to blow secondary air in domestic wood stoves. Reliable research data is unavailable to evaluate effectiveness of the various designs.

GOOD WOODS

Table 9-2 lists the most popular woods found in the United States. Each group is arranged in order of their descending fuel value. Looking in the left-hand column, you will see that live oak is better, for example, then black locust. Black locust, in turn, is better than dogwood.

Aside from their fuel values, there are certain other characteristics that users of woodburning appliances should be aware. For example, the bark of white birch gives off much soot while burning and is therefore prized more for its beauty than for its value as a fuel. All of the oaks have excellent fuel value, as does black walnut. All of these woods give off a pronounced acrid odor that may be unpleasant—especially if burned in an open fireplace or Franklin stove. Other types are apt to harbor insects and should be used up before warmer spring weather.

All evergreens (with the possible exception of red cedar and Douglas fir) should be used only in a pinch. The softer hardwoods such as basswood, yellow poplar, butternut, bit tooth aspen, and

Table 9-2. Relative Heat Values of Wood. Each Group is Arranged in Order of Descending Fuel Value.

Best	Okay	Only in a Pinch
Live Oak	Holly	Hemlock
Shagbark hickory	Pond pine	Catalpa
Black locust	Loblolly pine	Yellow poplar
Dogwood	Shortleaf pine	White spruce
Hophornbeam	Red maple	Black willow
Persimmon	Cherry	Bit tooth aspen
Apple	American elm	Butternut
White oak	Black gum	White pine
Honey locust	Sycamore	Balsam fir
Black birch	Douglas fir	Cottonwood
Red oak	Pitch pine	Basswood
Sugar maple	Sassafras	
American beech	Magnolia	
Yellow birch	Red cedar	
Longleaf pine	Bald cypress	
White ash	Chestnut	
Black walnut		

cottonwood are not as satisfactory as the denser kinds such as maple, oak, hickory, beech, yellow birch, and the various fruitwoods including apple, wild cherry, and the like.

GATHERING FIREWOOD

If you have a woodlot consisting of good hardwood trees, then you're more fortunate than most as firewood is becoming a valuable commodity. If not, there are still other sources to obtain free firewood, free for the cutting. For example, the National Forest Service around my home usually opens certain areas for the cutting of firewood for personal use. You cannot sell it, but you are entitled to cut a certain amount for your own use as fuel. In some areas, however, this privilege has been abused, and the Forest Service has closed these areas from public cutting. You'll therefore want to check with your Forest Service office closest you to make sure that these areas are open. Then find out exactly where they are located and what markings will be used to identify the areas.

Another source is right-of-way clearance for power lines or highways. When such construction is taking place, the power company or the highway department will usually be happy for you to cut all the wood you want in the areas to be cleared. This gives you free firewood and saves them some labor in the clearing.

You may also try some of the local farmers in your area. Many will be glad to sell you firewood cheaply—especially if you will do the cutting and hauling. The national forests are your best bet, provided you live within a reasonable distance from one. If you would like to find out more information about the national forest closest to you, and the possibility of cutting firewood in designated areas, write one of the following regional offices that is within your area.

If you live in Montana, northeastern Washington, northern Idaho, western North Dakota, or northwestern South Dakota, write to National Forest Service, Federal Building, Missoula, MT 59801. If you live in Colorado, Kansas, Nebraska, South Dakota, or Wyoming, then write to Federal Center Building 85, Denver, CO 80225.

For residents of Arizona and New Mexico, the Forest Service is located at 517 Gold Ave., SW, Albuquerque, NM 87101. Those people living in Utah, southern Idaho, western Wyoming, or Nevada should write to Forest Service Building, Ogden, UT 84403.

Residents of California should write to 630 Sansome St., San Francisco, CA 94111. Oregon and Washington residents should write to 729 NE Oregon St., Portland, OR 97212.

Residents living in Maine, New Hampshire, Vermont, Massachusetts, Connecticut, Rhode Island, New York, Pennsylvania, New Jersey, Delaware, Maryland, West Virginia, Virginia, and Kentucky should write to 6816 Market St., Upper Darby, PA 19082.

Residents living in Alabama, Arkansas, Florida, Georgia, Louisiana, Mississippi, North Carolina, Oklahoma, South Carolina, Tennessee, and Texas should write to 50 7th St. NE, Atlanta, GA 30323.

Residents of Illinois, Indiana, Iowa, Michigan, Minnesota, Missouri, North Dakota, Ohio, and Wisconsin write to 710 North 6th St., Milwaukee, WI 53203. Alaska residents should write to P.O. Box 1631, Juneau, AL 99801.

In all cases, write to the regional office of the National Forest Service and state that you would like to inquire about the possibility of cutting firewood in national forests if any cutting sites are currently available. They will inform you as to the availability of such lands, where they are located, the markings to look for when cutting and any other information that may be needed.

You may also try calling your local power company and/or highway department. Inquire as to the possibility of helping them clear right-of-ways in exchange for firewood.

If you're going to cut your own, try to arrange to do this in the early spring when the weed growth is not too great. The work will be easier. The wood will have all summer and early fall to season—plenty of time for the curing of most woods.

If you live near a river or stream, driftwood is always available—especially in late spring after the high water level has subsided. Driftwood comes in all types along river banks that is suitable for almost any use. Small, dry sticks are fine for starting a fire, while larger, moister ones are good for banking a fire.

You may have to do some scouting around, but look on the outside bank where the river or creek bends sharply. Chances are you can find a winter's supply of fuel in any one of these locations. The difficult part may be hauling it out. Unless there is a suitable road directly to the location, you'll have to either carry it out on foot or else use a boat to haul it to an access point. All this may be more trouble than it's worth. On the other hand, if you have a home or weekend cabin directly on the waterfront, the water will do the work for you—delivering it right to your front door.

Using driftwood for fuel does have its drawback. For example, driftwood from any source is likely to be contaminated with sand and silt. If much sawing or cutting has to be done on this wood, your saw or ax blade will be dull very quickly. Another disadvantage of driftwood—especially those pieces gathered from salt water— is the carrying of corrosive chemicals into the stove. Depending on the type and thickness of metal used in your stove, this corrosive material can cause portions of your stove to rust out in a much quicker time than it normally would. Still, in some circumstances driftwood can be a very satisfactory fuel.

PAP-O-FIRE

So you don't have any wood at all? You can build a pap-o-fire in your fireplace and still have lower heating bills, or so says Dwight Mills of Fredericksburg, Virginia.

Mills, now retired, has worked in the heating fields most of his life and has devised a neat, simple method for transforming old newspapers and used crankcase oil into "free" heat for the home. Generally, you take old newspapers and roll them up until they are about 4 inches in diameter. They are then held together with a single piece of reinforced packaging tape. These paper logs are then stood up in a tin can in which used motor oil has been poured, so that each roll of paper soaks up the oil.

These oil-soaked paper logs will give out good heat just like they are, but for more efficient burning Mills has designed a special two-level grate to provide sufficient air for complete, clean combustion of the oiled newspapers. For greater heat radiation, Mills uses three or four steel logs (or cylinders) on the grate's upper level, in addition to three oil soaked paper logs. He also uses one real wooden log to complete the combination. As the fire burns, the steel logs become hotter, which provides the air entering the fire to be preheated before it begins combustion. The result is a hot, clean, nearly pollution-free blaze that leaves practically no ash residue.

If you have a source of free used crankcase oil, like at the corner service station, this may be a way to make your own fuel logs for winter heating. The total cost of the pap-o-fire setup is less than $100 if you want to buy the system from Mills. For more information, you can write to Dwight Mills, 1500 Winchester Street, Fredericksburg, VA 22401.

You may be able to find newspaper rollers in many of the inexpensive gift catalogs. These devices cost only a few dollars and will enable you to consistently make tightly packed paper logs for use as fuel in your fireplace, or to aid in starting fires. Most of these kits come with their own supply of tie wire to keep the rolled logs tightly bound.

HARVESTING WOOD

Let's assume that you have found your supply of firewood. Now you have the problem of getting the wood from tree to stove in a safe manner. Your best bet, and often the desire of the land owners, is to look for downed trees—those that have been downed by high winds, ice storms, and the like. This makes the chore much easier and helps to clean up the woodlot.

The first step in cutting up a downed tree is to cut off the limbs approximately flush with the trunk. This operation is called *limbing*. A sharp ax is about the only tool that you will need, except on the largest trees where a chain saw or bow saw will make the work a lot easier and faster. Start on the free limbs (those not supporting the weight of the tree), working from the base of the trunk toward the top. The first cut on each limb should be made on the side facing the base of the trunk—the underside of the limb when the tree was upright. Cut and trim all of the free limbs as close as possible to the trunk. Protruding stubs can cause injuries and also make it more difficult to handle bucking stage.

Now comes the tricky part—the cutting of the limbs that are supporting the trunk. A wrong cut in the wrong place can cause the tree to come crashing down to the ground, pinning your ax or chain saw with it. The object is to lighten the trunk as much as possible before cutting too many of the supporting limbs. Depending upon the type of tree and several other factors, I usually start at the top end of the trunk when cutting the supporting limbs. The tree is first topped (the very end, thin branches are cut away). The next couple of supporting limbs are chosen and cut—starting the cut on the outside of the bend a few inches away from the trunk. When the limb is free, the stub is cut flush with the trunk. I then cut back or cut the trunk into sections after measuring for the correct length for my wood stove. The trunk is cut in sections back to the next supporting limbs. Again, a few more supporting limbs are chosen and cut away from the trunk. At this point, however, you must be careful to judge the role of each supporting limb as far as its support is concerned. The wrong cut and the entire trunk can turn and roll on you or knock your chain saw into your leg.

When making such cuts, make certain that you have adequate room to retreat should the tree start to roll or fall. Be on guard at all times during the cut; be ready to jump aside if the tree starts to roll. Be prepared to pull the chain saw away from the cut if you feel the weight of the tree pinching the guide bar. Even then it is wise to have an ax or second saw ready in case the saw binds during the cutting; a second cut further from the trunk will usually release a bound tool. Leave a few of the supporting branches to support the trunk during bucking.

Cutting the trunk and major branches into stove-length sections can be done with either a chain saw or bow saw. I've even sectioned some trees with an ax, but this is both time-consuming and hard work. After a little experience, most people can "eyeball" the sections and cut to the right size without measuring. At first it is best to measure each section, and then mark by scoring with a light saw cut or ax. Again, working from the tip of the tree toward the base, cut the trunk into sections as marked. Cut the remaining supporting branches as you get to them. The major branches that will be cut into firewood can usually be laid across the trunk or another section of wood and sawed on the spot. A sawbuck will make the job a lot easier.

In bucking the trunk, the position of the piece being cut will dictate your first cut. If the length is free (not supported by a branch) and not extremely large, I usually start on the top side of the section and cut right through with the chain saw. If the section is relatively large, I first make an undercut about one-fourth the thickness of the section and then finish the cut from the topside. When doing so, however, be careful not to let the section drop on a toe or foot.

If the section is supported by a branch, the cutting procedure is reversed. Make the first cut from the top, halfway through the thickness of the section. Then finish the cut from the bottom. If the cut is made completely through from the topside, the supported piece can buckle inward and pinch or bind the saw.

When all of the supporting limbs have been removed, you may still have a section of trunk to cut up. Since no supporting limbs are present, the trunk will normally be resting directly on the ground. If you try cutting through the trunk in this position, you're sure to cut into a rock on the bottom side and dull your saw blade. Rather, make all cuts (of the proper length) about halfway through the thickness of the trunk. Then roll the trunk over and finish these cuts from the opposite side.

The sections of wood are now ready to be hauled to your home or place of utilization. Try to pick a place around your home to store the wood so that it won't be unsightly. Pick a place close enough so you won't have to walk so far to carry in your firewood for the stove or fireplace. I also like to do the splitting near the storage location; this again cuts down on walking and carrying to a minimum.

SPLITTING WOOD

Splitting firewood can be fun if you have the proper tools and go about the task correctly. It can also be one of the most back-breaking experiences if you don't have the right tools or don't know what you're doing.

Green wood splits much more easily than seasoned wood, so it's a good idea to split the wood as soon as possible after cutting it. To do so, first separate the various pieces into size groups. Usually, any piece under 6 inches in diameter need not be split; it will dry relatively fast without being split. Due to its small size, handling is no problem. Six to 8-inch diameter pieces should be set aside to be split with an ax. Those pieces larger than 8 inches will normally have to be split with a *splitting maul*—sometimes used in combination with a *wedge*. Extremely large or difficult to split pieces (such as *locust*) may require the use of a *sledge hammer* and two or more wedges. A power splitter will do the job also, but it is an expensive tool.

When splitting, provide some support for the piece being split and to prevent your ax from striking the ground. A medium size log will be all that you need. Make sure you have the proper size handle on your ax, and know the feel of it during the swing. A handle that is too short for you can let the blade come back against your foot or leg.

The piece to be split should rest against the support log (*chopping block*) in a vertical position. If the log has been lying around for a while, the ends of the log will have started drying out. When the log is bucked, you will have one "new" or greener end on each piece to be split. This "new" end should be facing upright for splitting; the section will split easier in this position. In soft ground, you may also want to place a board under the other end to keep it off the ground. The ground, if too soft, will absorb much of the energy of the ax swing. If the ax should cut into the dirt, it will become dull very quickly.

If the piece is relatively small, I aim the ax blade at about the center of the section and let go. I give the ax handle a slight twist to rotate the blade the moment the blade comes into contact with the wood. This technique takes a little practice. Look for radial cracks on the larger pieces and aim for them. Take out small sections at a time if necessary. The cracks will aid you in making a clean split.

You can expect to get your ax stuck—especially at first. You'll try to take off too big a hunk, and your ax will bind in the split as your blow won't entirely cut through the piece of wood. When this happens, pound a wedge in the crack to release the ax. While you're at it, you might as well finish splitting the wood with the wedge.

You can avoid binding your ax by not cutting large pieces of wood directly in half. First, split away the edges until you have reduced the size of the piece to a reasonable size; then split through the remaining hunk.

In an emergency, like when your chain saw is not operating properly, you can cut some firewood for immediate use with the ax alone. You can fell the tree by chopping and then limbing the tree as described previously. Rather than chopping each piece of firewood in the desired length, use a maul and several wedges to split the entire trunk lengthwise. If the tree is relatively small, one split will usually be sufficient. On larger trees, however, you will want to first split the trunk in half and then into quarters, or possibly eighths. With the cross section area reduced in size, the individual pieces will then be easier to chop into usable lengths. This is not the recommended practice, since bucking a trunk with a saw is much easier. In certain cases, though, this could mean the difference between staying warm or being cold.

STACKING AND SEASONING WOOD

With your supply of wood cut and split into the proper-sized pieces, it must be stacked in a way to allow for seasoning. The time required for seasoning wood varies according to the species, but from six to nine months is usually sufficient to dry most woods.

Fig. 9-1. This piece of wood has been cut for only two months. It shows signs of cracking already.

Fig. 9-2. This piece of wood has been cut for over a year. It is well seasoned.

Softwoods such as pine should not be stored for much longer than a year or 18 months. They will begin to decay after this time and lose much of their heating value (Figs. 9-1 and 9-2). Hardwoods such as oak, hickory, and the like will normally last up to two years of storage before they start to decay. When you are gathering your firewood, don't have more than a two-year supply on hand at any one time. But I don't think most of us have to worry about having too much wood; it takes all of my energy just to cut enough to last one season.

Many people merely stack their firewood haphazardly in a pile and forget about it (Fig. 9-3). For the wood to season to the ideal dryness, though, it should be off the ground and sheltered from the weather (Fig. 9-4). Two 12-foot 2 × 4s laid in parallel on the ground will act as a support, as will parallel poles. When using wood, the poles should be treated with a wood preservative to prevent rot. With the wood elevated above the ground, air can circulate freely under the pile and speed drying.

Wood is normally stacked in two different ways—parallel and crisscross. If you're looking for the fastest drying time, then choose the crisscross method. For saving space, the parallel method of stacking is the way to go. Most stacks of wood should not be any higher than 4 or 5 feet. If the stack is much more than 2 or 3 feet, you will have to brace the ends.

Regardless of the method of stacking, the pieces of wood should be positioned so that the bark side is up to repel water—even if the wood is stacked under cover. Some rain will more than likely blow into a lean-to or similar shelter and spray the wood. If the wood is completely under cover, such as in a garage, then I can

Fig. 9-3. This is how the majority of firewood is stored—on the ground. The plastic drop cloth will prevent rain from soaking into the logs, but the layer of logs on the bottom absorbs moisture from the ground.

see no difference in the way the wood is stacked—barkside up or down.

WOODSHEDS

One of the simplest woodsheds is to drive four stakes into the ground (two at each end) and provide parallel poles between them for the wood to lay on. To protect the wood from the weather, you can spread a sheet of plastic material over the pile. Anchor it in place with rocks or other pieces of wood.

A little more sophisticated storage idea is to make end frames. Since these ends are movable, you may make your woodpile as large or small as you like. Merely move the ends inward or outward to accommodate the amount of wood. The weight of the logs on the bottom 2 × 4s of the end frames anchors the frames at the ends of the stack. When the wood diminishes, you just move the end frame in closer. Stack more wood on it to anchor it and get the remaining wood tightly stacked.

Figures 9-5 through 9-7 show ways of storing wood. You should be able to find one that suits your own needs.

PRESERVING YOUR WOODLAND

Many people living on farms or in rural areas will have access to their own woodlots. If managed properly, they can have a good supply of wood for fuel in years to come. Wherever farming has invaded former areas of timber areas, remnants of second-growth forests usually occupy the areas where the soil is least suited for cultivation. The value of these woodlands is often very high, but frequently their importance as shelter to crops and as sources of fuel, posts and poles are not fully realized. A little care in protection from fire, removal of insect-infested or diseased trees, prevention of injury to seedlings by overgrazing, and intelligent selection of trees to be cut to relieve crowding would mean more complete use of the soil. The quality of the timber produced would be improved.

Beneficial thinnings can often be made at the time of ordinary utility cuttings by selecting the trees good enough for the purpose desired, rather than the best trees of the stand. The selection of the thriftier, more promising trees for preservation is made easier by a simple classification used by foresters.

When crowding begins in a stand of trees of the same age, some trees secure an advantage in the competition for soil nutrients and light. These trees become taller than the ones with larger crowns and trunks. The largest trees are called *dominants*. They are the more rapid growers, the most vigorous of which would survive to maturity if natural competition proceeded undisturbed. The trees which are somewhat shorter than the dominants but still reach into the upper crown canopy are called *codominants*. These trees are

Fig. 9-4. A crude but effective shelter like this will insure that you have dry firewood when you need it.

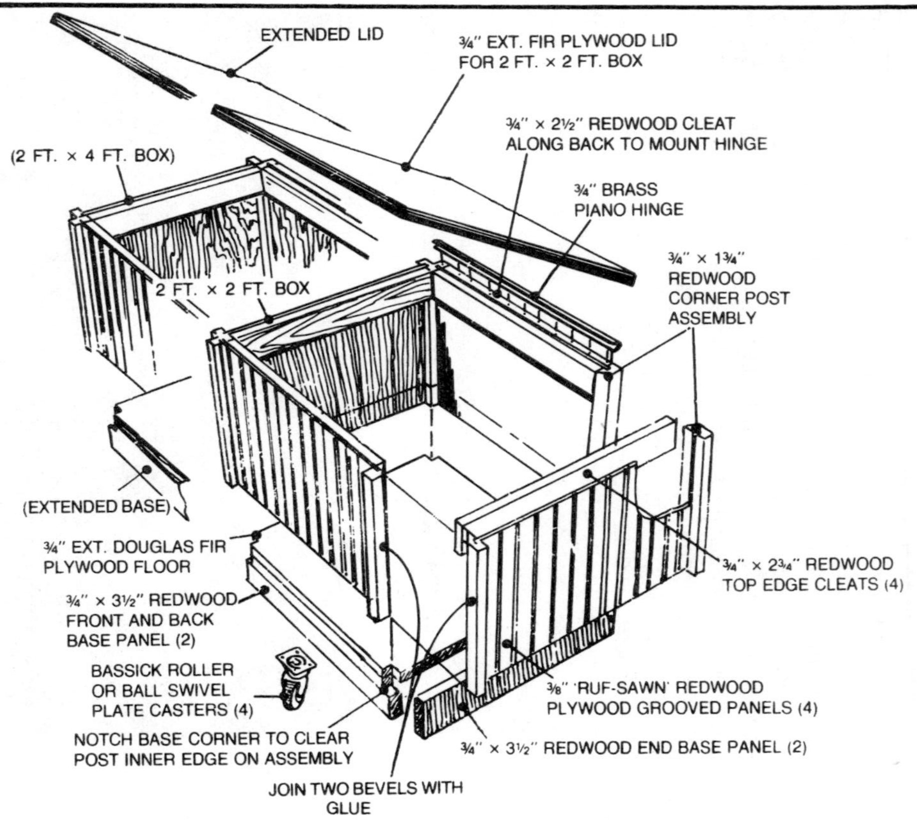
EXTENDED LID
¾" EXT. FIR PLYWOOD LID
FOR 2 FT. × 2 FT. BOX
¾" × 2½" REDWOOD CLEAT
ALONG BACK TO MOUNT HINGE
(2 FT. × 4 FT. BOX)
¾" BRASS
PIANO HINGE
¾" × 1¾"
REDWOOD
CORNER POST
ASSEMBLY
2 FT. × 2 FT. BOX
(EXTENDED BASE)
¾" EXT. DOUGLAS FIR
PLYWOOD FLOOR
¾" × 2¾" REDWOOD
TOP EDGE CLEATS (4)
¾" × 3½" REDWOOD
FRONT AND BACK
BASE PANEL (2)
BASSICK ROLLER
OR BALL SWIVEL
PLATE CASTERS (4)
⅜" 'RUF-SAWN' REDWOOD
PLYWOOD GROOVED PANELS (4)
NOTCH BASE CORNER TO CLEAR
POST INNER EDGE ON ASSEMBLY
¾" × 3½" REDWOOD END BASE PANEL (2)
JOIN TWO BEVELS WITH
GLUE

NOTE: ALL LUMBER SIZES ARE GIVEN IN NOMINAL DIMENSIONS. SIZES AS SHOWN ON THE DRAWINGS ARE THE NET SIZES OF THE LUMBER. FOR EXAMPLE, CORNER POSTS ARE SPECIFIED IN MATERIALS LIST AS 1″ × 2″ WHILE ACTUAL NET SIZE IS ¾″ × 1¾″.

USE EXTERIOR WATERPROOF GLUE AND GALVANIZED, HOT-DIPPED ZINC OR ALUMINUM NAILS.

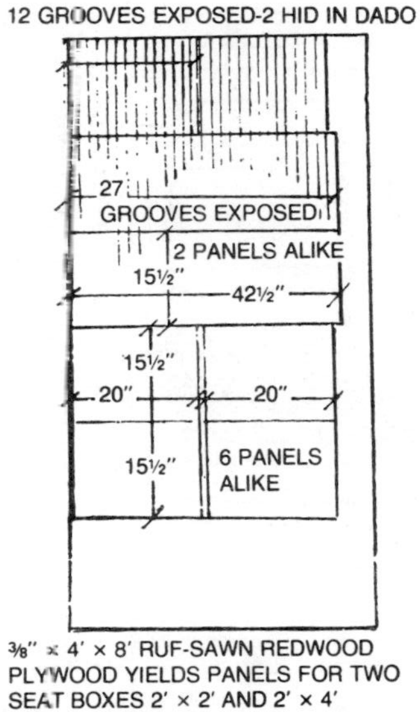

2′ × 2′ BOX

MATERIALS LIST

Part	Pieces	Size	Material
Bottom	1	20¾″ × 20¾″	¾″ Simpson exterior Douglas fir plywood
Front and Rear Base	2	20¾″	1″ × 4″ Simpson certified kiln-dried redwood
End Base	2	19¼″	1″ × 4″ Simpson certified kiln-dried redwood
Corner Posts	8	15½″	1″ × 2″ Simpson certified kiln-dried redwood
End Panels	2	15½″ × 20″	⅜″ Simpson Ruf-Sawn all-redwood plywood
Front and Back Panels	2	15½″ × 20″	⅜″ Simpson Ruf-Sawn all-redwood plywood
Top Edge Cleats	4	19¼″	1″ × 3″ Simpson certified kiln-dried redwood
Lid	1	23″ × 23″	¾″ Simpson exterior Douglas fir plywood
Piano Hinge	1	21″	Brass
Swivel Casters	4	2½″	

2′ × 4′ BOX

Part	Pieces	Size	Material
Bottom	1	20¾″ × 43½″	¾″ Simpson exterior Douglas fir plywood
Front and Rear Base	2	43½″	1″ × 4″ Simpson certified kiln-dried redwood
End Base	2	19¼″	1″ × 4″ Simpson certified kiln-dried redwood
Corner Posts	8	15½″	1″ × 2″ Simpson certified kiln-dried redwood
Front and Back Panels	2	15½″ × 42½″	⅜″ Simpson Ruf-Sawn all-redwood plywood
End Panels	2	15½″ × 20″	⅜″ Simpson Ruf-Sawn all-redwood plywood
Top Edge Cleats	2	41¾″	1″ × 3″ Simpson certified kiln-dried redwood
Top Edge Cleats	2	19¼″	1″ × 3″ Simpson certified kiln-dried redwood
Lid	1	23″ × 45¼″	¾″ Simpson exterior Douglas fir plywood
Piano Hinge	4	43½″	Brass
Swivel Casters	4	2½″	

Fig. 9-5. This attractive box is movable and can be used as a wood storage chest—either indoors or out. If made from redwood, it can be left to weather naturally, or the top, corners, base, and trim can be stained a dark shade, leaving the redwood panels light.

Fig. 9-6. A permanent, attractive wood storage area can be built. The concealed storage area is built into a fence and is quite compact. See Fig. 9-7 for construction details.

declining in vigor and would become overtopped and die if unrelieved by cutting. Intermediates are usually less than two-thirds the height of the dominants, with very small poorly developed crowns. Completely overtopped or understory trees are called *suppressed.* They have very small crowns and are on the verge of perishing.

In the natural struggle for existence, the weaker dominants decline to codominants. These in turn become intermediates. The intermediates become suppressed, and the suppressed trees die. If the stand is undisturbed, the understory trees never gain a place in the upper crown canopy. This unregulated competition often retards the growth of the entire stand.

The first thinning of trees usually becomes necessary when they are 30 to 40 years old. First consideration should be given to the removal of dead, diseased, broken, or poorly formed trees. When competition has not been too prolonged, thinning from above, or in the dominant stand, will greatly improve the vigor of the understory trees. If large material is needed and the dominants have attained suitable size, this method of cutting often coincides with practical necessity. Care should be taken in felling not to injure the smaller trees to be retained. If smaller material can be used and larger trees are desired later, the thinning should be made from below. Remove enough of the weaker suppressed and intermediate trees, and perhaps some codominant trees, to improve spacing and relieve the dominants from competition.

As a precaution against destruction of the entire stand by uncontrolled fires, slash from the cutting should be made into fuel or

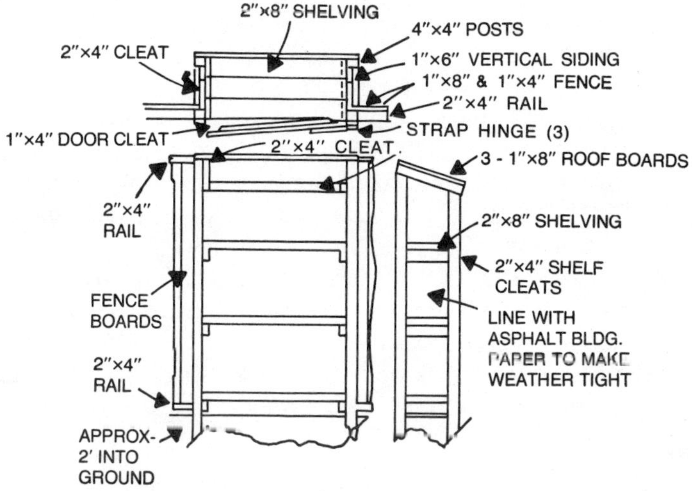

Fig. 9-7. For the concealed storage cabinet, 4 × 4 posts should be anchored in concrete because of the added strain to which they will be subjected. It would not be difficult to add a door onto the service area side, making the cabinet even more accessible. The cabinet roof is given a downward slant, so that it does not interrupt the top line of the fence. Asphalt paper can be laid over the 1 × 8 roof boards to keep the roof weatherproof. Make sure hardware and fastenings resist corrosion. More importantly, make sure that you specify all-heartwood redwood. The natural chemicals in the heartwood of redwood will protect your fence against decay and termite attack. No further treatment of the wood is necessary.

piled and burned so as not to injure the remaining trees. This may also prevent development in the slash of insects, which under certain conditions emerge from the limbs and tops and attack standing trees. Care should be taken not to permit the fire to spread through the stand and destroy the litter and leaf mold. The feeding roots of trees are usually very near the surface. Removal of the litter permits drying of the soil to a sufficient depth to kill the rootlets and retard growth.

COMMON TERMS

Sapwood is the outer portion, and ***heartwood*** is the center portion of the cross section of a log. The sapwood is usually lighter in color than the heartwood.

Springwood is the thin-walled, relatively weak, and usually light-colored portion of the ring. Summerwood is the thick-walled, heavy, and relatively strong portion of the ring that grows annually in the tree.

The term *softwood* is applied to the group of trees that bears cones. Examples are pine and spruce. *Hardwood* is applied to *broad-leaved* trees such as *oak* and *maple.* Actually, some softwoods are harder than some hardwoods.

Dense wood is heavy and hard; hickory is an example. The term "heavy" is used in defining the actual wood substance apart from any moisture it may contain. Even a light wood may have much weight if it is saturated with water. Heavy woods have high fuel value. Seasoning wood increases its fuel value.

The ***heat value*** of 1 cord of heavy wood, 1½ cords of medium-heavy wood, or 2 cords of light wood is approximately equivalent to that of a ton of coal.

Softwoods generally burn more readily than hardwoods; light woods burn more readily than heavy woods.

FACTS ON WOOD

■ Wood shrinks when it dries and expands when it takes up moisture.

■ Hardwoods shrink more than softwoods; heavy woods shrink more than light woods.

■ Boards from small logs tend to warp more in drying than boards of the same width from large logs.

■ Properly kiln-dried wood is as strong as air-dried wood.

■ The amount of sap in trees does not vary with the season of the year.

■ Wood in contact with the ground or kept moist for any length of time decays.

■ Wood continually submerged in water or kept perfectly dry does not decay.

■ Green fence posts set in the ground last as long as seasoned posts.

■ Heartwood is generally more durable than sapwood of the same species.

■ Pitchy wood is more durable than nonpitchy wood of the same species.

■ Dark woods are usually more durable than light woods.

■ The time of year when trees are cut has no great effect upon their resistance to decay if they are properly cared for after cutting.

■ Wood is not effectively protected from decay by painting.

■ Posts or trees with the bark on decay more rapidly than peeled posts or timbers.

■ Heavy wood is generally hard and strong; light wood is soft and weak.

■ Heavy wood holds nails well, but tends to split in nailing.

■ Seasoned wood holds nails better than green wood.

■ Seasoned wood is stronger than green wood except in shock resistance.

■ The growth change from sapwood to heartwood in the wood of the tree does not affect the strength.

■ Timber of the same species grown in widely different localities generally varies no more in strength than timber of the same species grown in one locality.

■ Woods of light weight are better temperature insulators than heavy woods.

■ Woods with prominent alternate annual growth rings of soft spring wood and hard summer wood are more difficult to work than woods which are either uniformly soft or hard.

■ Softwoods generally require less care in gluing than heavy woods.

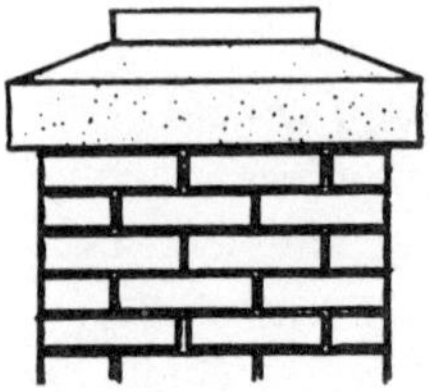

Chapter 10 Building and Tending a Fire

The type of stove or furnace you may have—regardless of its quality or efficiency—will be no better than the way you fire it. Several factors govern efficient heating with wood: type of wood, the amount the wood has been seasoned, the percent of "fill" you allow in the firebox, and others. This chapter shows how different types of kindling are used and how to arrange wood for starting a fire.

KINDLING

If you depend on wood heat, you will need plenty of good, dry kindling to carry you through the winter months. The first place to start is at your woodpile where you have been chopping and splitting your winter supply of firewood. In the early autumn months, the ground will not be frozen. This is the time to start picking up wood chips to use as kindling during the winter months. Make sure you gather the chips early enough. When the ground starts to freeze, they are extremely difficult, if not impossible, to pick up. More than likely you'll have to wait until spring before they can be gathered.

Wood Chips

You'll still probably be cutting and splitting to some extent all winter. Gather up the chips at the end of the splitting. The wood chips will likely supply you with at least 50 percent of the kindling that you'll require during the winter months.

These wood chips will have to be kept dry, which means stored off the ground under a shelter. A bushel basket makes a good storage container. It will allow air to circulate through the chips, dry them out, and keep them that way until they are used. Just make sure the basket is kept off the ground, preferably set on wooden blocks. If convenient, one basket of chips should be kept inside the house, near the stove, to season quicker (although chips will normally dry out in a very short time since they are so thin) and to insure that they will be dry when needed. Wood chips when stored under a shelter outside will sometimes take on some moisture under certain conditions. While these may still be used to start fires, the process will be quicker and easier if they are completely dry.

Scrap Lumber

The next most convenient source of kindling is possibly your scrap lumber bin—if you're a do-it-yourselfer who works on woodworking projects. This shouldn't be a very large supply, but maybe you can scrounge enough kindling to last a few weeks. You might try your local lumberyard or cabinet shop for a possible source of wood/lumber scraps to be used as kindling. Again, these wood scraps should be kept dry—the same as for wood chips—so the proper storage is compulsory.

Corncobs

Many farmers living in the country use corncobs to start their wood fires. When the grain is shelled from them to be used for livestock feed, the empty cobs are piled in a position so that air can circulate through them. This will help to remove any moisture left in them and also to keep them dry until they are used.

You can make corncobs even easier to light by filling a tin can with about an inch of kerosene. Place several corncobs (stacked upright) into the can, and leave them stand over night. The corncobs will draw up the kerosene the same as an ink blotter, and two or three of these kerosene-soaked cobs will easily start a fire if used as explained later.

Other Possibilities

If your lawn or woodlot boasts several large trees, chances are that many dead limbs and twigs will break from the trees each year and fall to the ground. These small limbs and twigs, when broken into pieces that will fit your firebox, make excellent kindling if stored in a dry place. Don't overlook the possibility of using pine cones and tree bark.

When you have to produce your own kindling, use dried softwoods (pine, spruce, *balsam,* etc.) and cut them into sticks about ¾ inch in diameter. Red cedar, either green or dried, may also be used for kindling when split into very narrow pieces. You might also want to try using a half-gallon waxed cardboard milk container. Cut or otherwise open the top of the milk container so that the full cross sectional area is accessible. Tightly pack balls of newspaper in the milk container until it is full. Light the open end of the container with a match, set it in the stove, and pile pieces of wood around the burning container. By the time the container burns out, you should have a roaring fire provided you have used dry wood that is not very large.

In the past, many users of woodburning stoves insisted on using ***kerosene*** to start fires. Do not use kerosene unless nothing else is available.

If you're building fires in a relatively new wood stove, only small fires should be made the first few weeks. The cast iron and firebricks must cure before larger fires can be built, or both materials may crack. If your stove does not have an ashpit in the bottom, such as in a metal drum stove, insulate the bottom of the stove with about 3 inches of sand. This will prevent the bottom from getting too hot. I've seen coals get so hot that they eventually melted through the metal bottom and fell out on the floor beneath the stove.

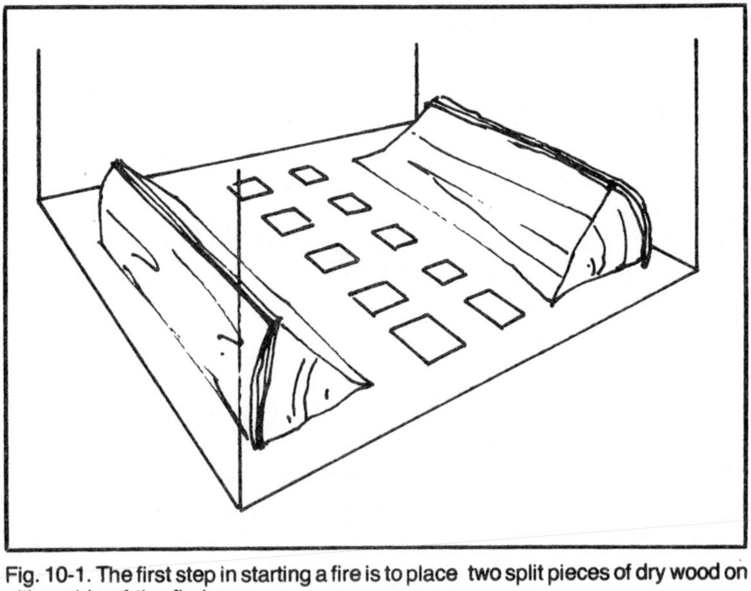

Fig. 10-1. The first step in starting a fire is to place two split pieces of dry wood on either side of the firebox.

Fig. 10-2. Crumple up newspapers into balls about the size of grapefruit. Line the floor of the firebox with them.

STARTING THE FIRE

The first requirement of good fire starting techniques is to have dry wood and kindling. Once the fire gets going, you'll be able to get by with wood that isn't so dry. The fire will quickly dry the wood that is placed into the firebox. I'm talking about "dry" as far as wetness is concerned—not about whether the wood is seasoned or not.

To start the fire, place two split pieces of dry wood on either side of the firebox (Fig. 10-1). There should be a 6-inch space between them. Crumple up newspapers—one sheet at a time—into balls and line the floor of the firebox (between the two splits of dry wood) with them (Fig. 10-2). Cover the balls of newspaper with dry kindling, laid either in tepee or log cabin fashion (Fig. 10-3). This kindling should be placed close enough together, so that the flames can move easily from one piece of kindling to another. Don't pack this kindling too tight, as the fire will be suffocated and die out before it catches. Place a few small pieces of split wood on top of the kindling. Be sure not to compress the kindling too tightly on the crumpled newspaper.

Open the draft just enough to encourage the fire without blowing it out, and yet enough so you won't fill your house with smoke as you are lighting the fire (Fig. 10-4). Light the paper and close the stove door (Fig. 10-5). You should hear the fire catching, and as it draws in more air through the draft it should burn more intensely. Once the stove is drawing well, add as much wood as the situation calls for.

As the fire burns, heat will be reflected between the two logs on either side of the firebox, keeping the logs burning and temperatures high. Stoves vary in the amount of wood they can burn, so you'll have to experiment to see just what your stove will do.

Fig. 10-3. Cover the balls of newspaper with dry kindling, laid in either tepee or log cabin fashion.

When the first logs have burned to coals, open the firebox and add more logs. If you're burning green wood, mix it about half and half with seasoned wood for best results. If you're using one of the Scandinavian front to back burners, rake the coals to the front of the stove before adding any new wood to the fire.

Once the fire is going good, leave the draft open for about 30 minutes to burn off most of the gases that cause creosote. Then partially close the damper to prevent the fire from getting hotter than necessary.

On some types of stoves, the draft inlets are from the tops of the stoves. The wood sizes should be reversed as previously described. The larger logs should be placed into the firebox first. Kindling is then added in a crosscrossed fashion, with the crumpled newspaper laid on top—being certain that most of it is located under the draft inlets. Light the fire and replace the lid, leaving the draft inlets open until the fire has been drawing well for about 15 to 20 minutes. Then control the draft for proper burn and temperatures.

HOLDING FIRES OVERNIGHT

The method used to hold a fire overnight depends upon the type of stove being used, as different stoves require different strategies. Generally, however, the following steps are necessary to hold a fire overnight.

- Let the fire die down in the evening so that only coals remain in the firebox at bedtime. You'll want enough coals to insure that they will ignite the overnight supply of wood, but not so many as to cause the fire to blaze up too soon. The coal bed may be adjusted by adding kindling if too weak, or by laying old magazines over the coal bed if too rich.
- Choosing the right type of wood for your stove is the hardest

step. You must experiment to find out what types of wood your stove requires for the best holding of a fire overnight. Remember that hardwoods hold longer than softwoods, logs in the "round" hold longer than splits, wood that is green holds longer than seasoned wood, and wood freshly brought in from the cold holds longer than wood that has been warmed by room heat prior to putting it on the fire.

■ Lay a split of half-dry hardwood on either side of the firebox and on the bed of coals—leaving a space of about 3 inches between them. Place a smaller piece of wood on the bed of coals between the two larger pieces. This latter piece should be relatively dry.

■ Lay paper over the wood foundation formed by the previous step. This paper should be glossy magazine paper about 10 pages thick. Overlap the various sections to form a good seal. The paper should form a seal open only at the ends.

■ Lay another split of wood on top of the papers that is relatively large and moist. If only dry wood is available, lay a piece of paper over this, too, so that it won't ignite too soon.

■ Shut the firebox door and close the damper. The fire is now banked.

If all goes well, the following morning the foundation logs will have been converted to glowing charcoal. The upper larger log will be dried out and ready for starting the daily fire. Open the damper, punch through the charred paper with a poker, and stir the fire

Fig. 10-4. Open the draft just enough to encourage the fire without blowing it out, and yet enough so you won't fill the room with smoke.

Fig. 10-5. Light the newspaper, close the door, and control the draft by sound by listening to the intensity of the fire as it burns. If you hear little or no sound of the fire catching, open the draft up wider. If you can hear the flames roaring, then close the damper slightly.

around a bit to break up the paper over the foundation. Clear the ashes away from the draft and open it. Shut the door, and the fire will now take off. Once the fire gets going, feed more wood as necessary. Let it burn for about 30 minutes with the draft wide open. This will help cut down on creosote buildup. Then close the damper to the desired position to maintain the required temperature.

TENDING A SMALL FIRE

In the early fall and spring when the weather is moderate, you'll want to keep smaller fires in your stove—usually in the early morning hours and perhaps in the evenings. You can use some kindling to form a bed of coals. Place two splits of green wood on either side of the firebox with the coals between them. As one stick burns, it should be replaced with another. Since the green wood will burn slowly, the heat will be very low. If more heat is desired, merely place a small piece of dry wood on the coals between the two pieces of green split wood. You can cut the fire down even more by using shorter pieces of wood in the firebox.

Chapter 11
Review of Wood Stove Designs

If you are seriously interested in supplementing an existing heating system or heating a home entirely with wood, a well-designed wood stove is far more efficient than a fireplace. It will extract more heat from each piece of wood.

The two basic features that distinguish most wood stoves on the market today are the method used to complete secondary combustion and the method used to control the rate of combustion. Complete combustion of volatiles depends upon how successfully secondary air is introduced to the stove without cooling the volatiles below 1100° F. At the same time, after a fire has passed the point where most moisture and volatiles have been withdrawn from a charge of wood, the fire may become hotter at an accelerating rate. If the burn is not controlled, frequent refueling will be necessary. The heating rate will be uneven, and warping, cracking, or other damage to the stove may result.

SIMPLE BOX STOVE

The simple *box stove* shown in Fig. 11-1 is the most basic and common design available. As its name implies, the design is usually embodied in a boxlike shape. Its operation principle is common to the familiar steel or cast-iron Franklin, pot bellied, and parlor stoves.

An adjustable intake is usually provided in the door or low on the front of the stove to admit primary air to support combustion. No provision is made for secondary air, so volatiles rise unburned from the charcoal bed and escape up the chimney. As a result, box stove

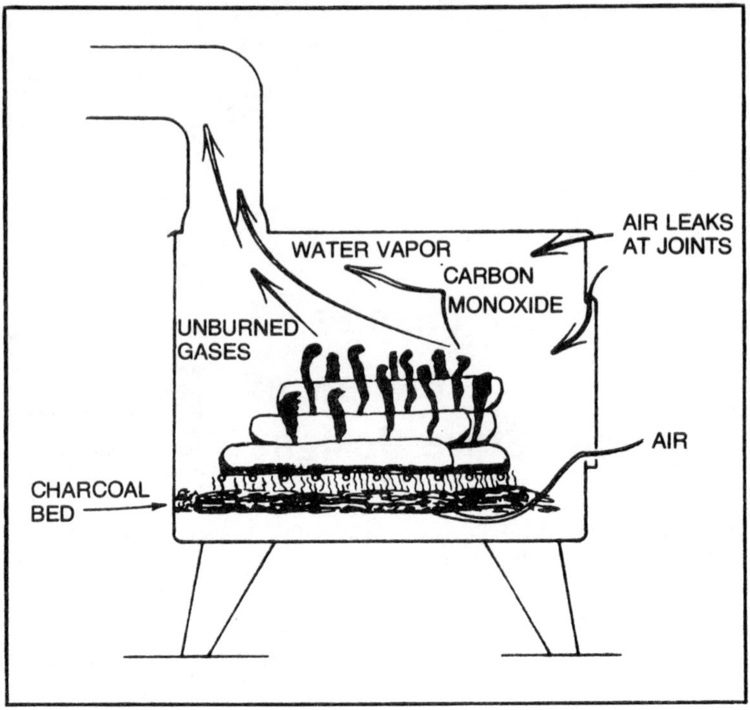

Fig. 11-1. Simple box stove design.

designs can be expected to deliver no more than 30 percent efficiency.

Generally, no control mechanism is provided in box stoves to prevent acceleration of combustion as the fire becomes hotter. Therefore, careful supervision of the fire is required to continually adjust the air intake and prevent uneven heating and rapid depletion of each fuel charge.

Even with careful attention, box stove designs are generally not tightly sealed, so it may be difficult to effectively control the burning rate. In the event of a chimney fire, excessive air leakage into the stove may prevent extinguishing the fire by cutting off its oxygen supply.

Simple box stove designs will usually be among the least expensive ones available. They offer better efficiency than conventional fireplaces, but the stoves generally burn more wood while delivering less heat than the airtight designs. A box stove may be well-suited when the stove will only be used occasionally, or when quaint, old-fashioned styling is more important than utility.

ASHLEY AIRTIGHT DESIGNS

Several companies manufacture good airtight stoves designed on the *Ashley* principle. Appearance varies. Some of these units may look like certain inexpensive box stoves, while others have external appearances similar to cabinet oil burners.

Ashley designs are usually constructed of sheet metal with gasket-lined doors and automatic air inlets. The fireboxes are lined with cast-iron panels or firebricks.

The Ashley design is shown in Fig. 11-2. Primary air is drawn into the stove through an automatic damper controlled by a thermostat. Air flows downward through a preheating channel on the outside of the firebox. It enters the firebox and passes across and upward through the fuel charge. Preheating of the primary air is desirable, since it helps maintain a higher firebox temperature and aids completion of the wood combustion cycle.

Most of the Ashley designs also provide secondary air inlets for use once the stove is in the charcoal burning stage. The usefulness of this feature depends on how the stove is operated. Charac-

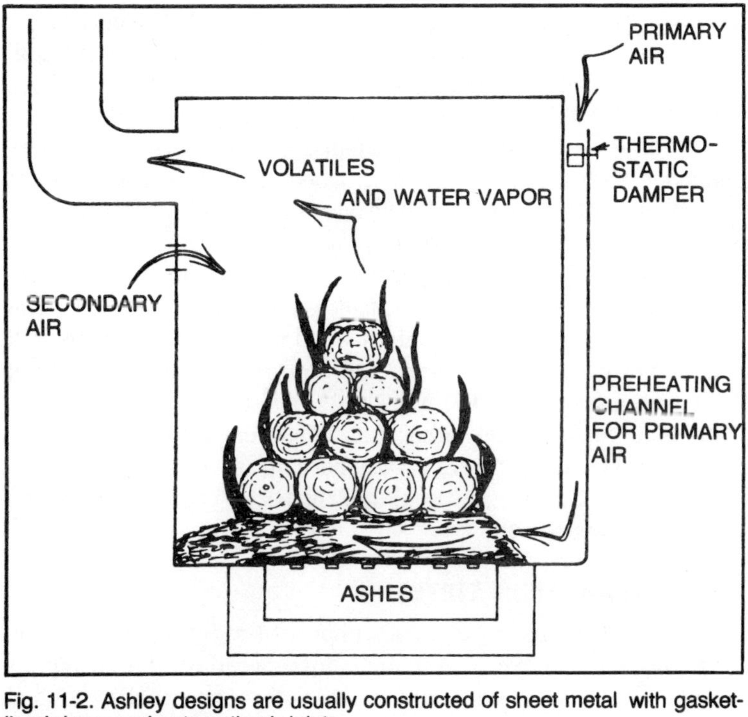

Fig. 11-2. Ashley designs are usually constructed of sheet metal with gasket-lined doors and automatic air inlets.

teristically, owners pack the stove full of logs at bedtime and turn the primary draft down. The firebox is chilled by the fresh fuel load and, with only enough primary air to keep the charcoal bed glowing, a low temperature burn begins so that volatiles are driven off well below their ignition temperature. Consequently, the volatile gases pass unburned into the flue despite secondary air. Nevertheless, even though the additional air contributes nothing to combustion, it may help in diluting the water vapor and volatiles so that condensation of acid and creosote is reduced at the expense of stove efficiency. If moderate temperatures are maintained, however, the stove should perform well. The Ashley designs have been used for many years and are the sources of a good, steady supply of heat.

BASE BURNING AIRTIGHT STOVES

Several designers have attempted to take the Ashley principle and add other design features to achieve more complete combustion. As early as 1832, a baseburning stove was developed. One type of base burning design is shown in Fig. 11-3. Base burners attempt to keep volatile gases hot enough to burn when mixed with secondary air coming through a preheating channel. The outlet from the firebox chamber is located at the base of the firebox, forcing all volatiles and other exhaust products to pass by the hot charcoal bed prior to leaving the stove. A base burning stove may tend to smoke when the door is opened for refueling. A base burner should be equipped with a bypass door that, when opened before refueling, allows smoke to go out the flue through the top of the firebox rather than the route followed during normal operation.

Control of the rate of burning in base burning stoves is provided either by a bimetallic mechanical thermostat or an electrical thermostat. As with most modern wood stoves, base burners will not always work as well as claimed. These units reportedly work best with a fairly high thermostat setting, and frequent refueling with small loads of fuel rather than infrequent large loadings.

Several manufacturers of base burning stoves also make similar furnaces, boilers, and multifuel furnaces capable of burning wood, coal, and/or oil.

DOWNDRAFT AIRTIGHT STOVES

In 1771 Benjamin Franklin improved stove efficiencies by inventing the "smoke consuming" stove based on the downdraft design. This system drives the volatiles through the charcoal bed rather than merely close to it. Dozens of downdraft designs have been attempted through the years, but successful implementation has been difficult. Franklin successfully operated the coal version of his downdraft stove, but smoking frequently occurred during refueling. Back puffing can be avoided by opening the draft wide a few minutes before refueling.

One type of downdraft stove is shown in Fig. 11-4. Primary air enters the stove above the fire, but below the main load of unburned logs. Primary draft air flows down and laterally through the glowing coal bed, drawing volatiles with it. Secondary air is introduced beneath the coals to burn the volatiles (over 1100° F). The heat of the fire is thus prevented from rising up through the load of logs, isolating wood in the upper region of the stove from the fire until it has dropped down into a burning position. Theoretically, even a full load of logs cannot cool the fire below. Volatiles may be released more slowly for effective burning.

A number of true downdraft stoves and furnaces have appeared on the American market since the time of Franklin, but none have survived due to many of the same difficulties that Franklin experienced. During the recent revival of interest in woodburning, several companies have reentered the market with new designs which claim to have solved the problems of bulky operation and back puffing.

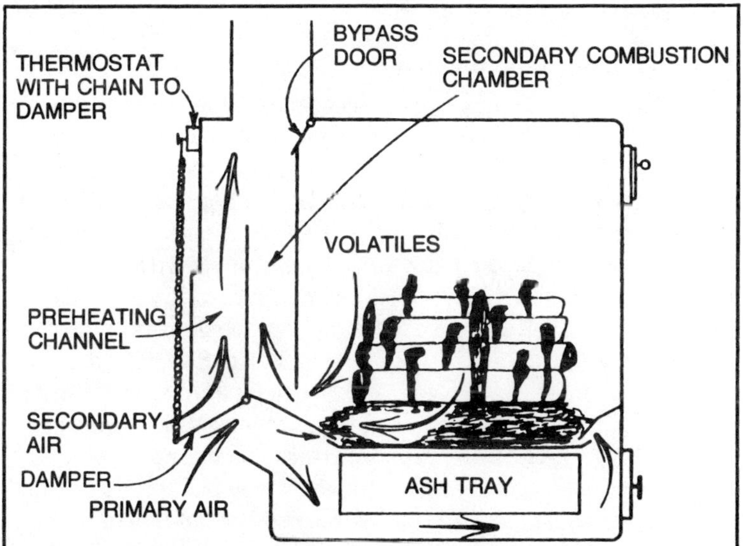

Fig. 11-3. Base burners attempt to keep volatile gases hot enough to burn when mixed with secondary air coming through a preheating channel.

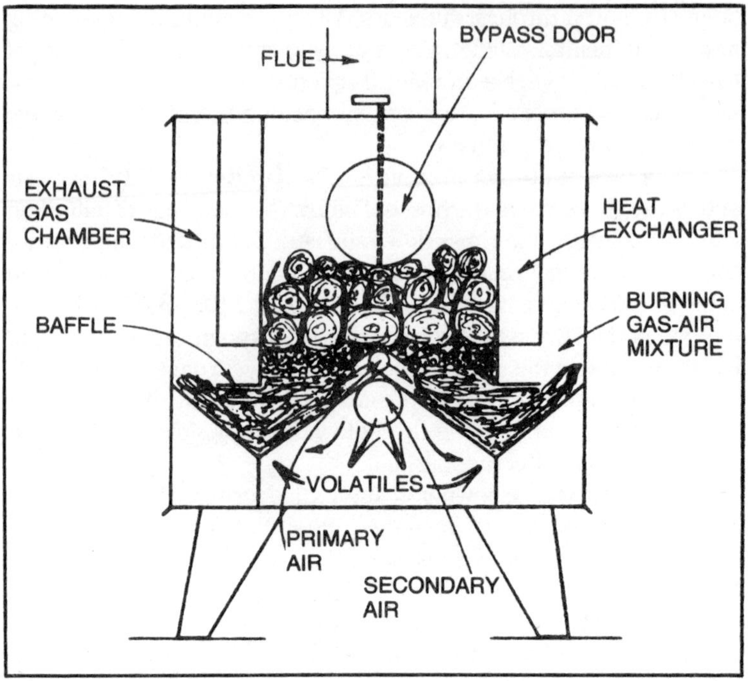

Fig. 11-4. In a downdraft stove, primary air enters the stove above the fire, but below the main load of unburned logs.

FRONT COMBUSTION STOVES

Front combustion stoves, sometimes called Norwegian or Scandinavian stoves, approach the problem of efficient wood burning differently than the designs discussed previously. Scandinavian designs rely on bringing draft from the end rather than from below, and burning the sticks evenly from one end to the other like a cigar. (Fig. 11-5). This is accomplished by means of airtight design and the addition of a baffle in the upper part of the firebox, forcing the draft into an S-shaped pattern. Air enters through the bottom of the door and contacts the embers in front of the fuel load. Exhaust gases and volatiles are forced to pass along the S-path forward around the end of the baffle, where they are mixed with secondary air and burned.

In theory, since the fire is uniform throughout a fuel cycle, even heat production is achieved once the draft is set. There is no need for an automatic damper. In practice, this may be only partially true. At the end of a fuel cycle in Scandinavian stoves, coals are raked toward the door to support fresh logs at the front end. The "cigar burn" may not be as pronounced as claimed, and rear portions of the logs may start burning ahead of schedule. The fire may tend to begin slowly, build to a peak, and then slowly subside to a few glowing coals over the course of eight or nine hours. Front combustion stoves work best when fueled with well-seasoned wood.

The front burning principle has been included in several recent American sheet metal stove designs. Most of the Scandinavian stoves marketed are imported and regarded by many as being tops in material and workmanship. These are constructed of heavy cast iron and are tightly sealed. Most are decorated with relief sculpture and porcelainized finish, including colors. Imported Scandinavian stoves are usually quite compact. They will fit on the hearth or install in most fireplace openings.

WOOD STOVE EFFICIENCY

Many manufacturers' claims are made about efficiencies of wood stoves. Actually, if you consider the relative efficiency of a stove compared to the conventional fireplace, there are a number of units whose relative efficiency is good. Reliable information on efficiency of specific stoves is not available. Efficiency tests are laborious and expensive, but because of renewed interest in wood-burning some reliable and impartial data should be available by the time you're reading this book. Some manufacturers have tested their own products, but self-testing may lead to questions about impartiality. Claims of efficiency above 50-60 percent are probably

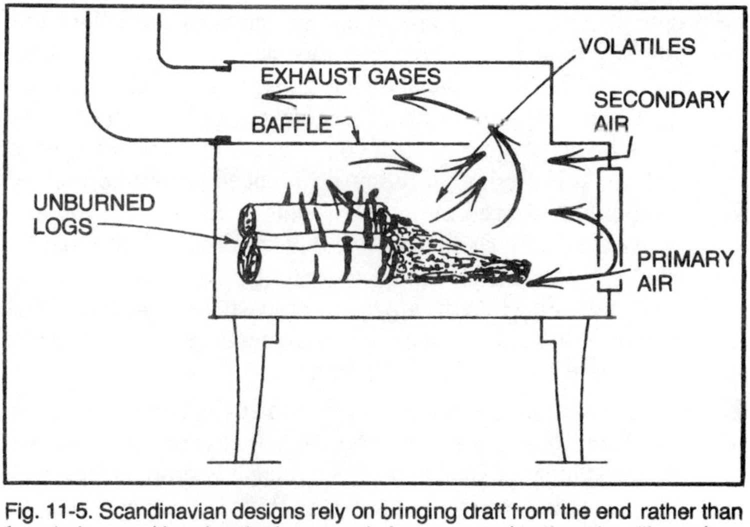

Fig. 11-5. Scandinavian designs rely on bringing draft from the end rather than from below, and burning the logs evenly from one end to the other like a cigar.

doubtful under household conditions, since careful packing of the firebox with small diameter sticks is essential for efficient burning. Large voids between sticks permit air to flow in stream lines right through the firebox without coming into contact with either solid wood or volatiles. The sticks also support combustion of one another through infrared radiation, so they should not be very far apart. Even wood-fired boilers used for the generation of electricity have been measured at efficiencies of only about 60 percent. This performance was achieved with an engineer in constant attendance, continuous stoking of wood chips, and the use of preheated forced air for both primary and secondary combustion.

For guidance in selecting a stove, the following rough efficiency ratings may be applied:

- Average conventional fireplace, 10 to 15 percent efficiency.
- Simple box stove (including Franklin fireplaces and antique stoves), 25 to 30 percent efficiency.
- Airtight wood stoves, 45 to 60 percent efficiency.

Top efficiency for each of these woodburning systems will be influenced most by the type of wood you burn in them, how well it has been allowed to season, how small it is split, and how tightly the wood can be packed in the firebox. As can be seen in Tables 9-1 and 9-2, there is a wide variation in the heat available from different species of wood. As much as 44 percent more heat can be obtained from wood by simply letting it air-dry before burning.

The overall efficiency of wood stoves will also depend partly on proper installation and operation. Selection of an undersized stove to heat a large area is a common mistake. Most owners will be tempted to operate the stove at high temperature to maintain room warmth, which will result in frequent refueling, and a very hot stove and chimney. On the other hand, if your stove is extremely large for the space to be heated, or if the heat produced cannot be rapidly dissipated or transferred into other rooms of the house, you are likely to have a room which is unbearably hot. You may attempt to limit the rate of burn to the point that the stove will become inefficient. Assuming that the stove selected will be large enough to heat most of your home, here are a few suggestions to assist you in locating the unit for the best efficiency:

- Most people want the living, dining, and kitchen areas to be the warmest. Therefore, place a large stove in or near the areas where you will spend most of your time. Even though bedrooms will tend to be cooler, some heat will be carried to them by convection (Fig. 11-6).

■ Place a large stove in a large room or a room interconnected to other rooms by large or numerous doorways. If the heater is placed in a small room with only one door, that room will likely be too warm. In such cases, it may be necessary to use portable fans to move air into other rooms, install a low-speed wall fan to rapidly dissipate air to an adjacent room or living area, install a grate with adjustable louvers in the ceiling to permit heat to rapidly rise to the overhead rooms, or remodel to develop a forced air system.

■ If your home has many small rooms which are not conveniently connected by doorways or halls, it may be best to install two or more smaller stoves.

■ Stoves installed in basements or on first-floor rooms will help warm the upstairs rooms. Lack of insulation between floor joists will facilitate rapid heat transfer. Opening stairway doors will also speed the heat movement upstairs.

WOOD STOVE INSTALLATION

Many people purchase woodburning stoves without considering the necessary installation steps. Running a single-wall smoke pipe through walls, ceilings, roofs, or windows does not meet local fire codes and may result in a serious fire or cancellation of fire insurance. Here are a few factors to consider.

—Provide a 36-inch clearance between the stove and any combustible surface or material to the front, back, sides, or above.

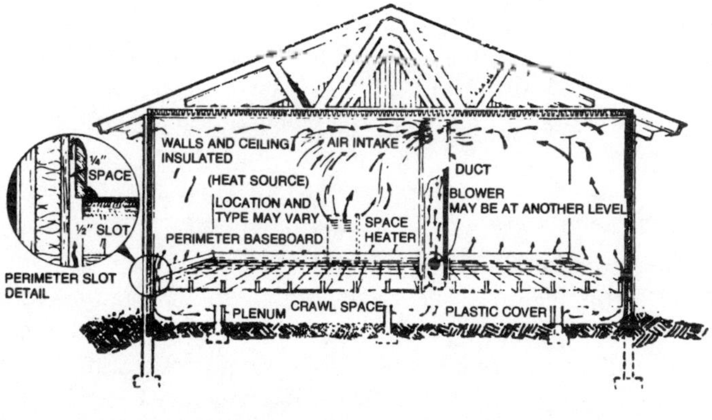

Fig. 11-6. Using an underfloor plenum with a slot around the perimeter of each room, you can heat an entire house with a single woodburning unit. Heat will be transferred by convection.

—Provide an 18-inch clearance between the stove and floor and a 24-gauge steel pad beneath the stove.
—Stove boards or hearths should extend 18 inches beyond the ash removal door, and 6 inches beyond the sides and back of the stove.
—Clearances may be reduced if combustible surfaces are protected with metal-covered stove boards, asbestos board, brick, tile, or other fire-resistant material.
—Protective materials should be installed with a 1-inch air space between the wall and material. One-inch lengths of ½-inch pipe may be used as nailing spacers between the protective shield and wall.
—Regularly check the stovepipe for corrosion and holes. Replace as needed.

Stovepipe should be used primarily to pass smoke from the stove through the flue to the outside. In earlier times, pipe was commonly run the full length of the room under the ceiling to provide maximum heating. Today, special heat exchangers are available for installation in the stovepipe to extract additional heat. These practices will cool the smoke and, in a cold room, often cause downdrafts back through the stove into the room. Remember, some heat is necessary in the pipe to create draft. Follow these guidelines for installing stovepipe:

—Install the small end of the pipe down or pointed toward the stove.
—Avoid long horizontal runs (length no more than three-fourths vertical high of the chimney where the pipe goes in), and provide an upward pitch of ½ inch rise for each foot of length. Be sure the pipe is securely supported.
—Minimize the number of elbows.
—The stove should be no closer than 36 inches to a ceiling wall or other combustible material. Allow at least 18 inches between the stovepipe and the ceiling or other combustible surface.
—Use at least 22-gauge pipe. Twenty-four gauge or heavier is desirable if available. Be sure the joints are tight and secured with two sheet metal screws to prevent accidental pipe separation.

The stovepipe should be cemented in the chimney masonry, or the metal thimble must be cemented to the masonry and the stovepipe tightly inserted into the thimble. Use high-temperature cement, making sure that the pipe does not protrude into the open flue area since draft will be reduced. If a masonry chimney is not used, the stovepipe must not pass through any floor, ceiling, or wall unless it passes through a ventilated metal thimble with a diameter at least 12 inches larger than the stovepipe.

A manual or automatic draft regulator or damper may be installed in the pipe in accordance with the manufacturer's instructions. Many airtight stoves do not require a damper, since draft is controlled either manually or automatically on the stove itself.

If you have an open fireplace and would like to install a wood stove in place of it, several approaches may be used. A few stoves are manufactured which have a shield built as an integral part of the stove to cover and attach to the fireplace opening. There is no such thing as a "standard size" fireplace opening, so special ordering is sometimes necessary. The entire fireplace opening can be sealed with a metal or asbestos board plate (Fig. 11-7). Some method must be provided for snugly attaching the shield to the fireplace so that no leaks occur into the opening. If leaks are present, draft may be reduced. There are likely to be whistling noises when the wind blows. If this method is used, be sure that the fireplace damper is securely locked open. While this method generally works well, should you ever decide to remove the stove and cover shield, you may find the inside of your fireplace very sooty.

The original fireplace damper may be removed and replaced with a plate or metal or asbestos board to seal the opening. If the damper opening permits, cut a hole large enough to run the stovepipe through it into the flue above the fireplace smoke shelf.

Fig. 11-7. This fireplace insert considerably improved the efficiency of the fireplace. When the blower of the furnace is running to circulate heat through the ductwork (the furnace burner is not running), this heater heats 1000 square feet of living space in all but the coldest weather.

Fig. 11-8. When your firewood gets damp, you can dry it faster if you will bring in a supply and store it near your wood stove.

Be sure that a tight fit is obtained; it may be necessary to weight the place or seal around the edges with stove cement. The pipe connection may be sealed by packing with fiber glass or asbestos rope. Your stove dealer can often suggest local machine or sheet metal shops which can provide custom shields or other installation accessories.

SUMMARY

Many woodburning stoves are available for use by the homeowner. New stoves are rapidly being introduced to keep up with the recent surge in demand. Few, if any, of the newest designs are really very different from some of the older designs. For the buyer seriously interested in heating with wood, these major factors should be considered:

- Select an airtight stove for efficiency from a dealer who can offer parts and service.
- Have an adequate supply of good quality, well-seasoned wood (Fig. 11-8).
- Be sure your stove is installed for safety.
- Have your insurance company clear the installation.
- Expect to work more for your heat than you did with gas, oil, or electricity.

Chapter 12
Coal Furnaces

People living in large cities in areas where wood for fuel is scarce (or too expensive to make its use worthwhile) may find that a coal-fired furnace is the way to save on your fuel bill. During the 1960s many homeowners had their coal-fired furnaces converted to oil. Oil, when compared to coal, was much cleaner and more convenient to use. Many of these homeowners are now having their furnaces converted back to coal or are installing new ones—trying to beat the rising cost of oil. While coal is certainly not as cheap as before, it will be cheaper than oil or electricity for heat. You can have the pleasure of knowing that you aren't depending on a foreign nation for our heat.

There are two types of coal furnaces—***hand-fired*** and***stoker***. A hand-fired furnace requires that coal be shoveled (by hand) into the firebox. When the supply burns up, more coal has to be shoveled in—the same as you would refuel a wood stove.

When a stoker is installed in a coal heating system, the fire usually takes care of itself. The less it is disturbed, the better it will be. Once the fire has been properly started and the correct cycle of operation established, there are only a few simple details that require attention at regular intervals.

Although stoker coals usually are supplied in uniform sizes, a certain amount of breakage always occurs in handling. Each load of coal should have at least a small quantity of fine particles. One thing to consider when filling the stoker is to shovel in equal parts of coarse and fine coal in the hopper. This will assure uniform feeding of the fine and coarse particles into the retort. Where the coal varies in size, this precaution should always be observed.

You will get more heat with less maintenance from a stoker-fired heating plant if you will inspect each shovelful of coal as it is placed into the hopper. It takes a little time, but by wearing gloves and inspecting each shovelful you are certain to turn up a spike, bolt, or other foreign object. If an object gets into the mechanism of the stoker, delay or breakage can result.

FIRING A STOKER FURNACE

Before firing a new stoker for the first time, make sure that all foreign matter (packing straps, bolts, crating, and packaging materials) is removed. In the fall, before firing the furnace for the season, again make sure the furnace is clean. If you are using a steam or hot water system, make certain that the water is at the proper level. Inspect the stoker to insure that it is properly serviced and lubricated, and be sure that the feed screw, windbox, and *tuyere* openings are free of ashes or other obstructions. If the hopper is fitted with a smoke pipe, check to see that it is clear. You are now ready to fire the stoker.

Fill the hopper about one-third full. Start the stoker and allow it to run until the retort fills level with the top. Then stop it and check all controls—both manually operated and automatic—to make sure they are working properly. Place several small pieces of dry kindling and twisted newspapers on top of the coal in the retort. Sprinkle a scoopful of coal over this and ignite the kindling. As the kindling is burning, start the stoker again. After making all the necessary adjustments, fill the hopper with coal.

The depth of fuel bed will vary depending upon the coking properties of the coal. Coals that form hard, strong coke usually require a deep fuel bed, while those that form soft coke generally burn better in a shallow fuel bed. The average coal furnace for home use, however, will require a fuel bed depth of about 8 inches for best results. Modern coal stokers usually are equipped with automatic air control devices and variable speed control for coal feeds. A correct air adjustment is one that will maintain a maximum depth of fuel bed without objectionable soot formation in the firebox and gas passages, or smoke discharge from the chimney. Once the stoker has been adjusted for a satisfactory grade and size of coal, it is best to use only that size and grade; in other words, avoid mixing different kinds of coal.

On stokers without a properly adjusted automatic air control device, one of the most common problems is the use of too much air. When too much air is supplied, the fire will become steadily thinner.

Ash will be blown clear of the fuel bed and will settle on the heating surfaces. When this occurs, a thin coke tree will appear at the center of the fuel bed. The tree sometimes is surrounded by dead, burned out areas adjacent to the edges which act as passages for the excess air.

An inadequate amount of air will cause a smoky fire. Also, if too much coal is fed into the firebox, the fire will be sluggish and smoky. Soot deposits will build up on the heating surfaces. The fuel bed will also form a heavy coke tree. The clinker "ring" will be thin, irregular, and may not form in the normal length of time. Large pieces of this coke tree will sometimes break away from the central mass and die out, resulting in an accumulation of unburned fuel around the sides of the firebox.

Another reliable and perhaps more sensitive indicator with most stoker coals is the nature and shape of the clinker formed by fusion of the ash and residues. With even a comparatively small amount of air in excess of requirements, the clinker tends to form deep in the firebed, and sometimes so low that it will partially obstruct the tuyere openings. The remedy for any of these troubles is to cut down the excess air or coal, as the case may be. Generally, it is best to adjust one or the other. Observe the results before making other changes. If too much air is indicated, cut down the air and note results before increasing the coal feed. Proceed in the same way where too much coal is indicated.

The amount of natural draft in a firebox is very important for efficient coal burning. Generally, the natural draft should be kept at the minimum. To check for the amount of natural draft, have the stoker running and the smoke pipe and chimney heated to normal operating temperatures. Open the fire door about 1 inch. Strike a match and hold it at the very edge of the cracked fire door. The natural draft should be just strong enough to draw in the flame gently toward the firebox. If the draft is excessively strong, set the smoke pipe damper to produce the correct draft as indicated by the match flame. If the installation is equipped with an automatic damper, the check damper should not be used. Instead, adjust the automatic damper. Many other conditions, both inside and outside the building, will affect the natural draft.

The clinker (ring of coke that forms in the firebox) should be removed at regular intervals using special clinker tongs. When removing the clinker, be careful so the firebed is disturbed as little as possible. To do this, grasp the clinker with the tongs. Raise it gently out of the bed of coals, turn it over, and shade off the loose ash and live coals. Rake the surrounding loose ash into the cavity with a curved poker, and level the fuel bed. In mild weather, run the stoker for a short period of time to recondition the fire after removing the clinker. The frequency of cleaning the fire depends on the ash content of the coal and the amount of coal consumed.

Most modern stokers have automatic controls designed to hold the fire (keep it from going out) during long intervals of mild weather when there is little or no demand for heat. The control promotes good combustion and converts ash into clinker. For this reason, it must be adjusted properly prior to use and rechecked annually before starting the furnace in the fall.

CHIMNEY INSPECTION

The chimney and smoke pipe should be checked at the start of each heating season. Unused stovepipe openings in the chimney should be checked and closed permanently. A corroded smoke pipe should be replaced, and care must be taken so that no air leaks exist at the check damper or where the pipe enters the chimney. Any of these openings, even though quite small, will act as miniature check dampers. If the coal furnace is of the warm air type (either gravity or forced air) and of cast-iron construction, the condition of the cemented joints should be checked periodically. Any leaks in these will permit gas and fly ash to enter the house. When recementing these joints, clean them thoroughly. Be sure that the new cement is spread to a uniform thickness so that when the parts are set together, there will be no opening or thin spots where the cement is likely to crack out shortly after the repair is made.

THERMOSTAT SETTING

The location of a thermostat for controlling the heat of a coal furnace in most homes should be carefully planned. It is extremely difficult to choose a location where the thermostat will not be affected by drafts and varying wall temperatures. A comfortable room temperature should be the guide in setting the thermostat rather than the reading on the dial. You can also provide a more comfortable atmosphere by adjusting register shutters for a more uniform heat distribution. It is also best not to vary the setting more than 10° between day and night setting of the thermostat. Wider variations throw a heavy load on the heating plant during the early morning warmup period. After a night at the low thermostat setting, the fire will not be in the best condition to deliver peak efficiency just when it is needed the most. When the fire has cooled down, (in the morning), however, this is the best time to remove the clinker and condition the fire before setting up the thermostat. The clinker will have had time to cool and harden.

TYPES OF COAL

While on the subject of coal-fired stokers, it's a good time to discuss the various types of coals used in this country as solid fuel. Coal may be classified in four main groups: *anthracite, semianthracite, semibituminous,* and *bituminous* coal.

Anthracite Coal

Anthracite coal contains from 93.31 to 100 percent of fixed carbon and from 0 to 7.69 percent of volatile hydrocarbons. It is rather hard to ignite and requires a strong draft for burning. It is quite hard, shiny, and is a grayish black. This type of coal burns almost with no smoke—giving it a certain advantage where smoke is objectionable.

Anthracite coal is known by different names, according to the size into which the lumps are broken. These names, with the generally accepted dimensions of the screens over and through which the lumps of coal will pass, are as follows:

- *Culm* passes through 3/16-inch round mesh.
- *Rice* passes over 3/16-inch mesh and through ¼-inch square mesh.
- *Buckwheat number 2* passes over ¼-inch mesh and through 5/16-inch mesh.
- *Buckwheat number 1* passes over 5/16-inch mesh and through ½-inch square mesh.
- *Pea* passes over ½-inch mesh and through ¾-inch square mesh.
- *Chestnut* passes over ¾-inch mesh and through 1-⅜-inch square mesh.
- *Stove* passes over 1-⅜-inch mesh and through 2 inch square mesh.
- *Egg* passes over 2-inch mesh and through 2-¾-inch square mesh.
- *Broken* passes over 2-¾-inch mesh and through 3½-inch square mesh.
- *Steamboat* passes over 3½-inch mesh and out of the screen.
- *Lump* passes over bars set from 3½ to 5 inches apart.

Semianthracite Coal

This type of coal contains from 87 to 93.5 percent of fixed carbon and from 7.69 to 12.5 percent of volatile hydrocarbons. It kindles easily, burns more freely than the true anthracite coal, and is highly sought as a fuel. It crumbles readily and may be distinguished from anthracite coal because it will soil the hand when just fractured, while anthracite will not. It burns with very little smoke. Semianthracite coal is broken into different sizes for the marker—the same as anthracite coal.

Semibituminous Coal

Semibituminous coal contains from 75 to 87.5 percent of fixed carbon and from 12.5 to 25 percent of volatile hydrocarbons. It differs from semianthracite coal only in having a smaller percentage of fixed carbon and more volatile hydrocarbons. Its physical properties are practically the same. Since semianthracite coal burns without the smoke and soot emitted by bituminous coal, it is a valuable fuel. Semibituminous and bituminous coals are known to the trade by the following names:

—*Lump coal* includes all coal passing over screen bars 1½ inches apart.
—*Nut coal* passes over bars ¾ inch apart and through bars 1½ inches apart.
—*Pea coal* passes over bars ⅜ inch apart and through ¾-inch openings.
—*Slack* includes all coal passing through bars ⅜-inch apart.

Bituminous Coal

This coal contains from 0 to 75 percent of fixed carbon and from 25 to 100 percent of volatile hydrocarbons. It may be divided into three classes. *Caking coal* is the name given to coals that, when burned in the furnace, swell and fuse together. They form a spongy mass that may cover the whole surface of the grate. These coals are difficult to burn, because the fusing prevents the air from passing freely through the bed of burning fuel. When caking coals are burned, the spongy mass must be frequently broken up with the slice bar in order to admit the air needed for its combustion. *Free burning coal* is a class of bituminous coal that is often called noncaking coal; the coals do not fuse together when burned in a furnace. *Cannel coal* is a grade of bituminous coal that is very rich in hydrocarbons. The large percentage of volatile matter makes it valuable for gas making, but it is seldom used for the generation of heat, except near the places where it is mined.

Lignite

Lignite or brown coal contains from 30 to 60 percent carbon, a small quantity of hydrocarbons, and a large amount of oxygen. It occupies a position between peat and bituminous coal, being probably of a later origin than the latter. This type of fuel is difficult to ship without breakage, and exposure to the weather causes it to absorb moisture rapidly. It is noncaking and yields a moderate heat. In this respect it is inferior to even the poorer grades of bituminous coal.

In comparing the heat value of coal to other solid fuels, 1 pound of good bituminous coal may be considered as the equivalent of 2 pounds of dry peat, 2½ pounds of dry wood, 3 pounds of dry bark, and 4 pounds of straw. Coal is the most efficient of the known solid fuels. It also can be converted to liquefied petroleum gas and to gasoline for use in automobiles, but the process is more costly than refining petroleum products.

IMPROVING COAL FURNACE EFFICIENCY

On older types of coal-fired furnaces that are not equipped with automatic air controls for limiting the amount of natural draft through the firebox, a relatively good control can be provided by automatically operating a check damper in the smoke pipe. The damper closes when the stoker starts to operate and opens when it stops, thus checking the draft.

The damper is operated by a motor unit of the type used with a wall thermostat for automatic damper control of hand-fired furnaces. For use with a stoker, the unit is wired to a double-pole contact switch instead of a thermostat. The switch has connections for three wires, the same as the thermostat, and is actuated by the armature relay of the stoker control.

The particular type of relay in the stoker control will determine the method of making and mounting the switch. An arm with dual contacts at one end can be mounted on the relay armature. When the armature moves in starting or stopping the stoker, it also moves the arm slightly to make or break the circuit between two spring contacts. The arm is insulated from the armature by a fiber grommet, and two brass contacts are insulated by being mounted on insulating blocks.

Most damper motor units have two electrical circuits, one of 110 volts to operate the motor and one of low voltage which flows through the thermostat to control the motor. The low voltage circuit consists of a three-wire cable (doorbell type). These wires are connected to the switch installed in the stoker control. The common or white wire is soldered to the arm of the switch, and the two remaining wires are connected to the two spring contacts in the relay or switch.

Mount the motor on a post, wall, or other convenient place. Connect one of its arms to the check damper by means of a small chain passed over pulleys. Plug the motor into a 120-volt circuit, and check its operation by starting and stopping the stoker. If the damper closes when it should open, reverse the connection of the two wires that are not used as the common, or change the position of the arm on the motor unit.

TROUBLESHOOTING COAL STOKERS

Assuming that your stoker has been properly installed and that the plant is of adequate capacity to heat the required areas in your home, the following information will help to locate common operating problems and enable you to correct them.

Problem: Excessive Smoke in Chimney

- Check the fuel bed as it is probably too thick. If so, reduce the coal feed.
- Not enough air may be causing the problem. Try increasing the amount of air to the firebox.
- Check for poor draft. Adjust the smokepipe damper if necessary.
- Coarse coal can cause excessive smoke in the chimney. Mix equal parts of fine and coarse coal.
- Check for a faulty chimney or smoke pipe, and repair if necessary.

Problem: Smoke in Furnace Room

- The firebox draft is probably too weak; correct by adjusting the smoke pipe damper.
- Check for too much air. If needed, decrease air by fan adjustment.

Problem: Not Enough Heat

- Coal feed rate is too low; increase the coal feed.
- Air supply rate may be too low; increase if necessary.
- Poor draft can cause the problem; adjust the smokepipe damper.
- Check for a dirty fuel bed; clean if necessary.
- Heating surfaces may be dirty; clean castings thoroughly.
- Coal may be too fine; mix equal parts of coarse and fine coal.
- Limit switch maybe set too low; check the manufacturer's instructions to correct adjustment.

Problem: Too Much Heat

- Check the controls for wrong setting; adjust to a comfortable setting.

■ Coal feed rate is too high; lower the rate of feed.

Problem: Soot Accumulation

■ Fuel bed is probably too thick; decrease coal feed.
■ Air supply rate may be too low; increase the amount of supply air.
■ Controls may be set incorrectly; increase operating periods of the "hold-fire" control.

Problem: Not Enough Air

■ Fan adjustment may be set too low; increase the fan opening.
■ Siftings may be in the windowbox; clean the windbox.
■ Remove any water in the airduct.
■ Retort openings may be blocked; remove foreign material.
■ Furnace room maybe too airtight; allow more air to be admitted into the area.

Problem: Hopper Gas

■ If you have an excessively dirty fuel bed; clean the bed thoroughly.
■ If the clinker is in the retort, remove it.
■ If the fuel bed is too thick, reduce the thickness of the bed.
■ If the hopper lid is not sealed, refit the lid and gasket.
■ If the hopper smoke pipe is plugged, remove the cap and clean the pipe.

Problem: Incorrectly Sized Coke Trees

■ Heavy trees indicate the feed is too fast; increase air or reduce coal feed.
■ Thin trees indicate too much air; increase the feed, decrease the air, or maintain a thicker fuel bed.
■ Coal may be too fine; mix coarse and fine coal.

Problem: Fly Ash

■ Fuel bed is too thin; increase the rate of coal feed.
■ Too much air will cause the problem; decrease air.
■ Clean a dirty fuel bed.

Problem: Fire Goes Out

■ Check the control circuit for operation, and then rebuild the fire.
■ If the clutch is out or a pin is sheared, throw in the clutch, renew the shear pin, or remove the obstruction.

- Coal may be arched in the hopper; break the arch.
- Fuel-air ratio is wrong; adjust one or the other.
- Reduce the firebox draft if there is excessive draft.
- If the drive belt is slipping, tighten or replace it.

Adequate, regular lubrication insures efficient operation of all movable parts, long equipment life, and minimum maintenance cost. Never lubricate too much. Observe the manufacturer's instructions when lubricating the various parts.

Chapter 13
Combustion and Miscellaneous Fuels

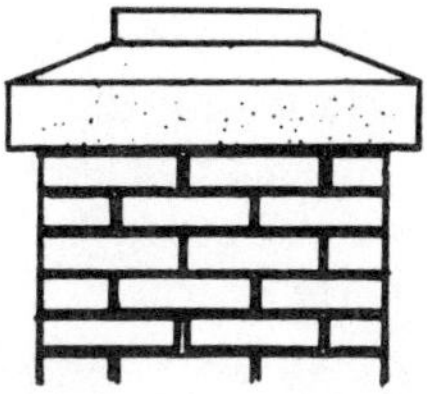

Combustion is the very rapid chemical combination of two or more elements accompanied by the production of light and heat. The atoms of some of the elements have a very great affinity or attraction for those of other elements. When they combine, they rush together with such rapidity and force that heat and light are produced. Oxygen, for example, has a great attraction for nearly all the other elements. For carbon, oxygen has a particular liking. Whenever these two elements come into contact at a sufficiently high temperature, they combine with great rapidity. The combustion of coal in the furnace is of this nature. The temperature of the furnace is raised by kindling the fire, and then the carbon of the coal begins to combine with oxygen taken from the air.

COMBUSTION

When carbon and oxygen combine, they form CO_2 or carbon dioxide. When hydrogen and oxygen combine, they form water or H_2O. These are called the products of combustion. The oxygen required for combustion is usually obtained from the air, which is a mixture composed of approximately 23 parts of oxygen and 77 parts of nitrogen by weight. The nitrogen that enters the furnace with the oxygen takes no part in the combustion. It passes through the furnace and up the chimney without any change in its nature.

When carbon is burned to carbon dioxide, one atom of carbon unites with two atoms of oxygen. Carbon has an atomic weight of 12. Oxygen has an atomic weight of 16, so the molecular weight of CO_2 is $(1 \times 2) + (2 \times 16) = 44$. You can see that CO_2 is composed of

12/44, or 27.27 percent of carbon, and 32/44, or 72.73 percent of oxygen. To burn a pound of carbon to CO_2, therefore, requires 32/12, or 2-2/3 pounds of oxygen. If oxygen is taken from the air, it will take 11.6 (2-2/3/.23) pounds of air to supply the 2-2/3 pounds of oxygen. This is because only 23 percent of air is oxygen. The combustion of a pound of carbon to CO_2 may be represented by equations, but 1 pound of carbon requires 11.6 pounds of air for complete combustion. Of this air, 2.67 pounds are oxygen. This oxygen combines with the pound of carbon, forming 3.67 pounds of carbon dioxide. The 8.93 pounds of nitrogen contained in the air pass off with the CO_2 as a product of combustion.

Let's consider the complete combustion of 1 pound of hydrogen. The product of the combustion is water, H_2O, which consists of 2 parts hydrogen to 16 parts oxygen. Therefore, 1 pound of hydrogen requires 16/2, or 8 pounds of oxygen to unite with it. The air required to furnish 8 pounds of oxygen is 8/.23, or 34.8 pounds.

The combustion of carbon may not be complete. If insufficient oxygen is supplied to the burning carbon, it is possible for the carbon and oxygen to form another gas, carbon monoxide, CO, instead of carbon dioxide, CO_2. The combustion of 1 pound of carbon to form CO requires only one-half the oxygen that would be necessary to form CO_2. This is because in CO gas, 1 atom of carbon to CO_2 requires 11.6 pounds of air.

The amount of heat, in Btu, developed by the complete combustion of 1 pound of fuel is termed the *calorific value* of that fuel. It is also sometimes called the heat value or the heat of combustion. It may be determined most accurately by burning a known weight of the fuel with oxygen in an instrument known as a *calorimeter.* The gases resulting from the combustion pass through a known weight of water and give up their heat to the water. By noting the rise of temperature of the water, it is possible to calculate the amount of heat absorbed. This will give the heat that would be produced by the combustion of 1 pound of the fuel.

MISCELLANEOUS FUELS

Coke is made from bituminous coal by driving off the volatile matter. It consists of from 88 to 95 percent of carbon, ¼ to 2 percent sulfur, and from 4 to 12 percent ash.

Wood is used for fuel in localities where it is plentiful. It contains from 20 to 50 percent of moisture when cut, and this percentage is not reduced much below 20 percent by drying. Wood has a calorific value of 6,000 to 7,000 Btu per pound.

Peat consists of vegetable matter that is partly carbonized and found at the surface of the earth. It contains from 75 to 80 percent water when cut and must be dried before it can be used as fuel.

Bagasse is the product left after juice has been extracted from sugar cane. It is used to some extent in tropical and semitropical countries. Naturally, its use is limited to the places where the sugar cane is grown.

Dried tan bark, straw, slabs, and sawdust are usually confined to tanneries, sawmills, and threshing outfits.

The combustible elements contained in oil fuels are the same as those in coal, namely carbon and hydrogen, and possibly a small proportion of sulfur. The heat of combustion per pound of oil, or the calorific value, may be found approximately from the chemical analysis of the oil by the same formula as that used for finding the heat value of coal. A more accurate method, however, is to burn a known weight of oil in a calorimeter and to measure the heat generated, from which the heat per pound of the oil may readily be calculated. From the results of available tests, the heat of combustion per pound of oil fuel varies between 17,000 and 21,000 Btu. The average calorific value of Texas and California crude oils seems to be about 18,600 Btu per pound.

HOW MUCH HEAT DO YOU NEED?

Regardless of the type of fuel you plan to use for heating your home, you'll have to find out just how much heat you need before choosing a heating system. For modern central heating systems, such calculations are usually left to engineers and contractors. There are zoning requirements to consider, sophisticated control systems and other technical problems to work out, and differences can cause price variations of over $1000. This is not so with most of the new wood and coal burning units. Many units are designed to allow do-it-yourself installations. You'll have to figure out your own heating requirements, so you can choose the most efficient and economical heater for your needs.

There are several ways to determine the heat required for an area. I cover only the more usable ones for homeowners.

In one method, heat losses from each room are first determined in Btu. (One Btu, or British thermal unit, is the heat necessary to raise 1 pound of water 1° F.) Before the heat loss can be determined, the following information must be secured about any room to be heated.

- The cubic feet in it.
- The square feet of glass surface.

■ The square feet of wall surface exposed to the air. From this, subtract the square feet of glass as determined in number 2 above. (Forget inside walls that adjoin heated rooms.)

■ The square feet of floors and ceilings exposed to the outside or that do not adjoin heated areas.

■ The temperature desired for the room. Good practice usually places this at 68 to 70° F.

You will also need to determine the minimum expected outside temperature for your area. Then proceed as follows. Multiply the cubic feet in the room by the temperature rise desired. Divide the quantity obtained by 56 (this is the number of cubic feet of air one Btu will raise 1° per hour). The result is the number of Btu required to raise the air in the room to the desired temperature.

Continue by multiplying the square feet of exposed glass surface by the temperature rise desired. Multiply this result by 1.25 (the Btu loss through 1 square foot of glass of single thickness per hour). If you have storm windows, use a factor of .62.

Multiply the net square feet of wall surface as found in the preceding paragraph by the temperature rise desired, and multiply this result by the factor shown in Table 13-1, which corresponds to the construction of the wall in question. This gives the Btu loss through this type of wall per hour.

CHOOSING A HEATING UNIT

Once the required heat is found in Btu for the area in question, you can easily choose an appropriate unit from one of the manufacturer's specifications.

Here's an example of how the calculations would go. Let's say you want to heat a 17.5 × 13-foot kitchen/dining area that has a ceiling height of 8 feet. To get the cubic feet of space to be heated, you multiply 17.5 × 13 × 8, which gives you 1820 cubic feet. The window over the sink measures 3.5 × 2 feet for a total of 7 square feet.

Assume there are three outside walls of 12 feet, 13 feet, and 17 feet. Adding these and multiplying the result by the ceiling height (8 feet) gives an outside wall area of 336 square feet, from which you must now subtract the 7 square feet of window space. This leaves you with 329 square feet of wall area. To summarize:

Room: 17.5 × 13 × 8 feet = 1820 cubic feet
Window: 3.5 × 2 feet = 7 square feet
Walls: 8 (12 + 13 + 17) − 7 = 329 square feet

Assume that this room is located in an area where the outside temperature drops no lower than 0° F, and the interior is to be heated to 70° F. There's a temperature difference of 70°.

$$\frac{1820 \text{ (cubic feet)} \times 70 \text{ (temperature difference)} =}{56}$$

2275 Btu needed to heat air

7 (square feet) × 70 × .62 (storm windows) = 303 Btu lost through glass

329 (square feet) × 70 × .25 = 5757.5 Btu lost through wall

2275 + 303 + 5757.5 (results above) = 8335.5 Btu

A total of 8335.5 Btu is required to raise the temperature of this room 70° when the outside temperature is 0° F. The smallest wood or coal heater will furnish the required Btu.

In the case of a wood or coal furnace used to heat your entire home, calculate all rooms in the home as just described. Add the total heat loss from each to get the Btu output required of a furnace to heat the home.

CHECKING HEATING EQUIPMENT

Another quick way to determine the required heat in Btu for a particular area is to check existing heating equipment of a different

Table 13-1. Heat Transmission Table. The Right-Hand Column Gives the Amount of Btu Transmitted per Square Foot, per Hour per Degree Difference in Temperature Between Inside and Outside Air.

solid brick wall	
4 inches thick	0.60
8 inches thick	0.42
concrete wall	
8 inches thick	0.48
12 inches thick	0.34
frame wall	
wood siding	0.48
same with insulation	0.34
same with sheathing and insulation	0.25
glass	
single window	1.25
storm window	0.62
ceilings	
Sheetrock or plaster	0.40
same with insulation	0.20
floors	
Double wood floor	0.24
Concrete on ground	0.40
Wood with crawl space	0.20

type. For example, if your living room is now heated with an electric baseboard heater, check the nameplate for its rating. Let's assume that you are now using 2000 watts of electric heat in your living room, and you want to determine the amount of Btu to use with a wood heater. Merely multiply the total watts by 3.4 or $2000 \times 3.4 =$ 6800 Btu. The number 3.4 is the conversion factor. When electric heat is utilized and you want to match your existing heat with wood or coal, multiply the existing wattage by 3.4. The answer will be in Btu.

Some homeowners are now using hot water baseboard heat, with the water heated by coils through a woodburning unit. A simple rule of thumb method for determining the hot water radiating surface required is to allow 1 square foot of radiating surface for each 2 square feet of glass surface, one for each 20 feet of outside walls, and one for each 200 cubic feet of room contents. In the room example just given, it would work out this way:

$$\frac{7 \text{ square feet of glass}}{2} = 3.5 \text{ square feet.}$$

$$\frac{329 \text{ square feet walls}}{20} = 16.45 \text{ square feet}$$

$$\frac{1820 \text{ cubic feet}}{200} = 9.10$$

Radiation requirement would then equal 29 square feet.

In these examples, to save space, it was assumed that the areas above and below the room were heated—and therefore there was no heat loss. For other cases, use the factors in Table 13-1. Multiply by the square feet and the temperature difference to obtain the heat loss.

Chapter 14
Questions and Answers

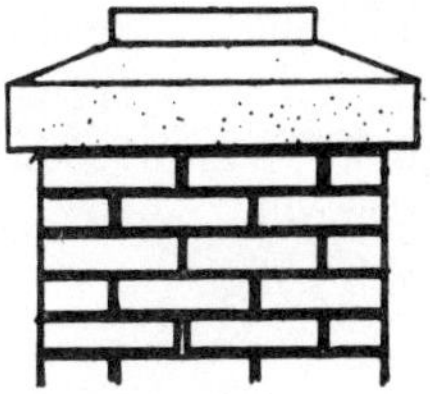

This chapter contains questions and answers concerning chimneys and heating.

CHIMNEYS AND DRAFT

Q. *What is the purpose of the breeching?*

A. It conveys the gases from the stove uptake to the chimney.

Q. *What is the purpose of the chimney?*

A. The chimney furnishes the draft necessary to burn the fuel and discharges the gases where they will not be a nuisance.

Q. *Of what materials are chimneys built?*

A. The materials are steel, brick, and reinforced concrete.

Q. *What is meant by natural draft?*

A. It is draft produced by a chimney without the aid of fans or blowers.

Q. *How does a chimney produce a draft?*

A. As the gases inside the chimney are hotter than the outside air, they weigh less per cubic foot. They tend to rise, being lighter. As they rise, the cooler, heavier air flows in through the ashpit and furnace to take their place. Thus, a continuous circulation is set up.

Q. *How is the draft pressure measured?*

A. It is measured usually in inches of water, with an inch of water corresponding to a pressure of .036 pound per square inch.

Q. *What is mechanical draft?*

A. It is a draft produced by mechanical means, as by a fan or blower.

Q. *What is forced draft?*

A. It is a form of mechanical draft in which the air is forced into the ashpit at a pressure greater than that of the atmosphere.

Q. *What is induced draft?*
A. It is a form of mechanical draft in which the gases are sucked out of the furnace setting and discharged into the chimney by a fan located between the furnace and the chimney.
Q. *What is balanced draft?*
A. It is a system of combined forced draft and induced draft in which the quantity of fuel fed, the air supplied, and the draft are all automatically kept in correct relation to one another. It makes no difference what the load on the furnace may be.
Q. *What are the advantages of forced draft?*
A. It facilitates the use of small sizes and low grades of fuel that cannot be burned successfully with natural draft. Forced draft enables high rates of combustion to be attained.

MEASUREMENT OF PRESSURE

Q. *How is steam pressure measured?*
A. It is measured usually in pounds per square inch.
Q. *What is meant by gauge pressure?*
A. It is the pressure indicated by the gauge on a boiler. It is the pressure of the steam above that of the atmosphere.
Q. *Does the atmosphere have pressure?*
A. Yes. At the sea level and in fair weather the atmosphere exerts a pressure of about 14.7 pounds per square inch.
Q. *What is a vacuum?*
A. It is a space that contains no matter of any kind—solid, liquid, or gaseous.
Q. *Does a vacuum have pressure?*
A. No. A true or perfect vacuum is a space devoid of all pressure.
Q. *What is a partial vacuum?*
A. It is a space in which the pressure is lower than that of the surrounding atmosphere.
Q. *How is the pressure in a partial vacuum measured?*
A. It is measured sometimes in pounds per square inch, absolute, and sometimes in inches of mercury.
Q. *What is meant by absolute pressure?*
A. It is the pressure measured above the zero point in a perfect vacuum.
Q. *How is the absolute pressure found?*
A. It is found usually by adding 14.7 pounds to the pressure indicated by the gauge. Absolute pressure is more accurately found by adding to the gauge pressure the pressure of the atmosphere as found by using a barometer.

Q. *What is a recording pressure gauge?*

A. It is a pressure gauge that has a pen instead of a pointer. The pen is supplied with ink and traces a line on a sheet of paper that is moved under it by clockwork. The pressure at any time is recorded on the moving sheet. Usually the sheet is a circular chart ruled with lines that enable the pressure indicated by the ink line to be read easily.

Q. *How are vacuum gauges usually graduated?*

A. Gauges show the vacuum in inches of mercury.

Q. *How is the reading of the vacuum gauge changed to absolute pressure?*

A. It is changed by multiplying the reading, in inches of mercury, by .49. Subtract the result from 14.7.

Q. *What does the factor .49 represent?*

A. It represents the pressure per square inch exerted by a column of mercury 1 inch high. It is also the pressure corresponding to 1 inch of mercury.

FUELS AND COMBUSTION

Q. *What fuels are used for heating?*

A. The fuels are solid, liquid, and gaseous.

Q. *What are the solid fuels?*

A. Solid fuels are coal, coke, lignite, peat, wood, bagasse, and others.

Q. *What are the various kinds of coal?*

A. They are anthracite, semianthracite, semibituminous, bituminous, and lignite.

Q. *How does anthracite differ from bituminous coal?*

A. It has less than 8 percent of volatile matter, whereas bituminous coal has from 25 to 50 percent of volatile matter.

Q. *What is volatile matter?*

A. It is that part of the coal that is driven off in the form of vapors and gases when the coal is heated.

Q. *Of what does volatile matter consist?*

A. It consists of gaseous compounds of carbon and hydrogen (hydrocarbons), sulfur, moisture, nitrogen, and carbon dioxide.

Q. *What remains after the volatile matter is driven off?*

A. Carbon and ash remain. The carbon is in the form of coke and is called *fixed carbon.*

Q. *Why is coal analyzed?*

A. Coal is analyzed to determine its composition.

Q. *What is a proximate analysis?*

A. It shows the percentages of carbon, hydrogen, oxygen, sulfur, and nitrogen.

Q. *What is meant by heating value?*

A. It is the heat, in British thermal units, developed by burning 1 pound of a fuel. It is also called the calorific value and the heat value.

Q. *What is combustion?*

A. It is burning, or the chemical union of combustible materials with oxygen, accompanied with the production of light and heat.

Q. *What combustible matter does coal contain?*

A. It contains mainly carbon and hydrogen, or compounds of these, and sometimes sulfur.

Q. *What is meant by complete combustion?*

A. It is the complete burning of the carbon, hydrogen, and sulfur in a fuel. The carbon burns to carbon dioxide, the hydrogen to steam, and the sulfur to sulfur dioxide.

Q. *What is essential for complete combustion?*

A. An ample supply of air is needed so that the carbon, hydrogen, and sulfur may unite with all the oxygen they can take up.

Q. *What is incomplete combustion?*

A. It is combustion in which the carbon does not get enough oxygen and is burned only partly, becoming carbon monoxide.

Q. *Why is incomplete combustion wasteful?*

A. It is wasteful because carbon burned to monoxide yields only 4,450 Btu per pound. Carbon burned to dioxide yields 14,600 Btu per pound. A pound of carbon burned to monoxide wastes 10,150 Btu.

Q. *Why are flue gases analyzed?*

A. The gases are analyzed to determine their composition and to judge the efficiency of combustion in the furnace.

Q. *What does the flue gas analysis usually comprise?*

A. The analysis shows the percentage of carbon dioxide in the flue gases, but to be accurate it ought also to show the percentage of carbon monoxide.

Q. *Why is the carbon dioxide reading usually sufficient?*

A. It indicates the percentage of excess air, which is the most common source of heat loss.

Q. *Why is the carbon monoxide reading of value?*

A. Even with a high percentage of carbon dioxide (indicating little excess air), there may be a considerable loss of heat by the escape of carbon monoxide, caused by an insufficient supply of oxygen in the air.

Q. *What causes smoke?*

A. Smoke is caused by not enough air, an improper mixture of air and fuel, too small a combustion chamber, and too low a temperature in the furnace.

Q. *At what temperature do the hot gases enter the chimney?*

A. The gases enter usually at 500° to 600° F., although the temperature may be higher or lower.

Q. *On what does the cross-sectional area of a chimney depend?*

A. The area depends on the quantity of coal to be burned per hour, as that determines the quantity of gases to be removed by the chimney.

Q. *Does a chimney for an oil-burning furnace need to be as large as for a coal-burning furnace of equal power?*

A. *No.* It may be smaller in height and in area, because oil burning requires less draft and produces a smaller quantity of gases.

Q. *What is the nature of fuel oil?*

A. It is the residue left after the lighter hydrocarbons have been distilled from crude petroleum.

Q. *How is oil prepared for use as a furnace fuel?*

A. Oil is heated to render it fluid, strained, and atomized by the action of a steam jet or by a mechanical device through which it is forced under pressure. The atomizing device is commonly called a burner.

Q. *How is oil fuel stored?*

A. It is stored in tanks of steel or reinforced concrete usually placed underground.

Q. *What kind of furnace is required for oil burning?*

A. The furnace must be lined with refractory material. In case the flame from the burner should snap out because of a slug of water in the oil, the heat of the lining will at once reignite the oil spray.

Q. *What is the most common method of atomizing the oil?*

A. The most common method is by a steam jet blowing across the mouth of the oil orifice. Air may be used in the same way.

Q. *How much steam is required for atomizing oil?*

A. The steam required is .3 to .5 pound per pound of oil.

Q. *What is the heating value of oil fuel?*

A. The value is from 17,000 to 21,000 Btu per pound, an average value being about 18,500 Btu.

Q. *What evaporation should be expected with oil fuel?*

A. The evaporation is 12 to 13 pounds of water per pound of oil, but under exceedingly good conditions an evaporation of 16 pounds can be obtained.

Q. *What is the flash point of an oil?*

A. It is the temperature at which the oil will begin to give off inflammable vapors.

Q. *What are the requirements of a good oil fuel?*

A. It may be either crude petroleum or fuel oil. Its flash point should not be below 140° F., the percentage of water not over 2 percent, and the sulfur not over 1 percent.

Q. *If oil should escape and become ignited, how should the fire be put out?*

A. Put out the fire by throwing sand or loose earth on it. Never use water, as that will only serve to scatter the burning oil.

Q. *How is a uniform pressure maintained on the oil that is being fed to a furnace?*

A. Pressure is maintained usually by a pump that draws the oil from the storage tank and forces it to the burners. A relief valve on the discharge line may be set to produce any desired pressure in the pipe. A pressure of from 40 to 80 pounds per square inch may be used.

Q. *How are fuel oils graded?*

A. They are graded according to their specific gravity as indicated by the readings of a Baumé hydrometer.

Q. *What is specific gravity?*

A. It is the ratio between the weight of a given volume of oil and the weight of an equal volume of water, both at a temperature of 60° F.

Q. *How is the specific gravity found from the Baumé reading?*

A. It is found by dividing 140 by the sum of 130 and the Baumé reading.

Q. *What is the specific gravity of an oil of 20°Baumé?*

A. According to the rule, the specific gravity is

$$\frac{140}{130+20} = .9333$$

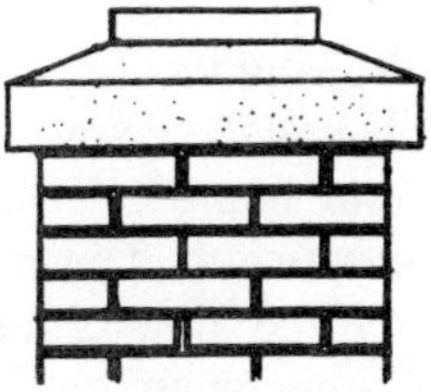

Chapter 15 Simplified Heating Calculations

When most modern comfort conditioning systems are designed for new buildings, engineers usually apply their knowledge and use of extensive calculations to determine the ideal system for the building—one that is adequate, dependable, and troublefree. It must have a reasonable installation cost and annual operating cost, and be easy to service and maintain. There are sophisticated control systems to work out, zoning requirements, and the like—not so with most woodburning units. You'll still need to know your heating requirements, but elaborate calculations aren't necessary.

STEPS IN DETERMINING HEAT LOSS

Basically, determining heat loss through walls, roof, ceiling, windows, and floors requires just four steps:

- Determine the net area in square feet.
- Find the proper heat loss factor from tables.
- Allow for infiltration.
- Multiply the area by the factor; the product will be expressed in Btu.

Heat loss calculations can be made more quickly and efficiently using a prepared form such as Table 15-1. With spaces provided for all necessary data and calculations in the proper order, the procedure becomes routine and simple. All that's required is that you include the correct information on the form and be certain that your calculations are accurate. The result is a room by room survey showing exactly how much heat will be required to maintain your home at a comfortable temperature during cold weather.

FLOOR PLAN

Note from Table 15-1 that the exact dimensions of all windows, doors, ceilings, floors, and the like are required. Make a floor plan of your home. Such a sketch will not only make identification of particular areas easier when you're using the calculation form, but it will also help to lay out the system components in some cases. So before getting into the actual calculations, let's briefly discuss how to lay out a floor plan.

Floor plans are not made full size because the size of the drawing would be impractically large. The drawing is reduced so that all the distances on it are smaller than the actual dimensions of your home or room, but all dimensions are reduced in the same proportion. The ratio or relation between the size of the drawing and the dimensions shown are the actual building dimensions, not the distance that is measured on the drawing.

The most common method of reducing dimensions in proportion is to choose a certain distance and let that distance represent 1 foot in the building. This distance is then divided into 12 parts; each one of these parts represents 1 inch. If ½-inch divisions are required, these twelfths are further subdivided. When measurements are laid out on the drawing, they are made with the reduced foot rule calibrated as previously described called an *architect's scale*. When a measurement is taken on the building itself, it is made with a standard rule or tape measure. See Fig. 15-1 for a practical example of the architect's scale.

Begin your sketch by first measuring the outside perimeter of your home, drawing the wall lines and inserting dimensions as you

Table 15-1. Heat Loss Estimating Form.

NAME OF PROJECT OR AREA ______________________
DESIGN CONDITIONS: Outside Dry Bulb Temperature ______________
Inside Dry Bulb Temperature ______ Temperature Difference ______

Part 1: Ceiling Type and Insulation Thickness ______________
Floor Type and Insulation ______________
Type of Windows ______________
Type of Walls and Insulation ______________

Part 2: Infiltration
area width × area length × ceiling height

$$\frac{(\quad)\text{ width} \times (\quad)\text{ length} \times (\quad)\text{ ceiling height}}{60}$$

Heat Loss
= ________ × 1.08 × ________ T.D. = ______

Part 3: Heat Loss

Windows:	________	square feet × 0.55 ×	________	T.D. = ________
Walls:	________	square feet × 0.13 ×	________	T.D. = ________
Roof:	________	square feet × 0.20 ×	________	T.D. = ________
Floor:	________	square feet × 0.09 ×	________	T.D. = ________

TOTAL HEAT LOSS IN Btu/h ________
Divide by 3.4 to obtain heat loss in watts ________

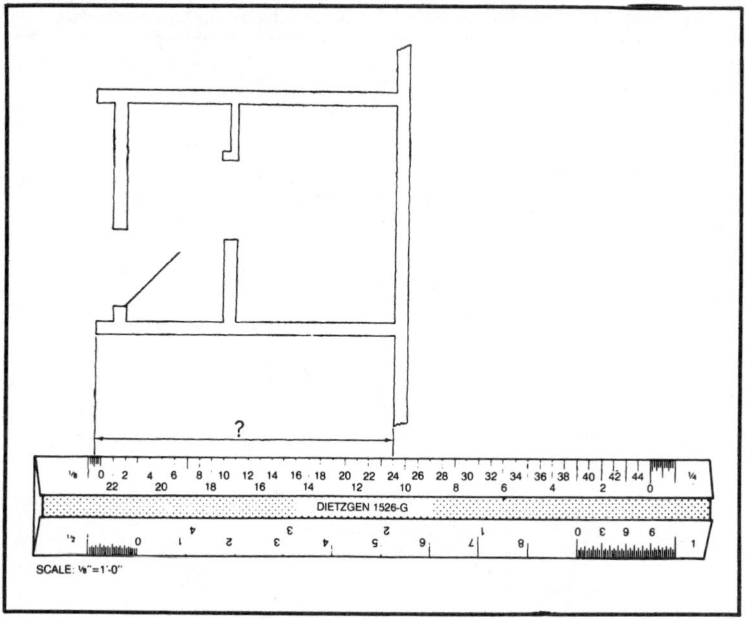

Fig. 15-1. Architect's scale.

go. Then measure and locate all windows and doors on the drawing along with chimneys, carports, etc. Continue by measuring all inside partitions, doors, etc., and indicate their locations on your sketch. When you have finished, your drawing should look something like Fig. 15-2.

Note on the drawing the location of storm windows and doors; the amount of insulation in your attic or above the finished ceiling, outside walls, and under your floor; the type of roof, floor, etc. The blank spaces on the calculation form in Table 15-1 will be a guide as to what information is required for a complete heat loss estimate. Try to obtain all this information before you begin your calculations; you will save time and come up with a more accurate estimate.

COMPLETING THE CALCULATIONS

The heat loss calculation form is based on design conditions of the home and the area in which the home is located. Outside design conditions are the extremes of temperature occurring in a specific locality, while the inside design conditions are the degrees of temperature and humidity that will give optimum comfort.

The outside dry-bulb temperatures for calculating heating loads are shown in Table 15-2. For cities and localities not listed in

Table 15-2. Outside Design Conditions for the United States and Canada.

State & City	Winter DB	Summer DB	Daily Range	Summer WB	Latitude Deg
ALABAMA					
Anniston	10	95	M	78	35
Birmingham	10	95	M	78	35
Gadsden	10	95	M	78	35
Mobile	20	90	L	80	30
Montgomery	20	95	M	78	30
Tuscaloosa	10	95	M	78	35
ALASKA					
Anchorage	-24	-	M	-	60
Barrow	-48	-	-	-	70
Bethel	-43	-			60
Cordova	-13	-			60
Fairbanks	-57	-	M	-	65
Juneau	-5	-	-	-	60
Ketchikan	4	-			55
Kodiak	4	-			55
Kotzebue	-46	-	-	-	65
Nome	-36	-	-	-	60
Seward	-4	-	-	-	60
Sitka	2	-	-	-	60
ARIZONA					
Bisbee	30	100	H	72	30
Flagstaff	-5	85	H	61	35
Globe	30	105	H	76	35
Nogales	30	105	H	72	30
Phoenix	35	105	H	76	35
Tucson	30	100	H	72	30
Winslow	-5	95	H	65	35
Yuma	40	110	H	78	35
ARKANSAS					
Bentonville	0	95	M	76	35
Fort Smith	5	95	M	76	35
Hot Springs	10	95	M	78	35
Little Rock	10	95	M	78	35
Pine Bluff	10	95	M	78	35
Texarkana	10	100	M	78	35
CALIFORNIA					
Bakersfield	30	105	H	70	35
El Centro	35	110	H	78	35
Eureka	30	90	M	65	40
Fresno	30	105	H	74	35
Long Beach	35	90	M	70	35
Los Angeles	40	90	L	70	35
Montague	15	95	M	70	40
Needles	25	115	H	–	35
Oakland	30	80	M	65	40
Pasadena	40	95	M	70	35
Red Bluff	15	100	H	70	40
Sacramento	30	100	H	72	40
San Bernardino	30	105	H	72	35
San Diego	45	80	L	68	35
San Francisco	35	80	M	65	40
San Jose	40	90	M	70	35
COLORADO					
Boulder	–15	95	M	64	40
Colorado Springs	–10	95	H	65	40
Denver	–10	95	H	64	40
Durango	–5	95	H	65	35
Fort Collins	–15	95	M	65	40
Grand Junction	–5	95	H	65	40
Leadville	–10	95	M	64	40
Pueblo	–15	95	H	65	40
CONNECTICUT					
Bridgeport	0	85	L	75	40
Hartford	0	90	M	75	40
New Haven	0	85	M	75	40
New London	5	85	L	75	40
Norwalk	0	85	L	75	40
Torrington	0	90	M	75	40
Waterbury	0	90	L	75	40
DELAWARE					
Dover	10	90	M	78	40
Milford	10	90	M	78	40
Wilmington	5	90	M	78	40
DIST. OF COLUMBIA					
Washington	10	90	M	78	40

State & City	Winter DB	Summer DB	Daily Range	Summer WB	Latitude Deg
FLORIDA					
Apalachicola	25	95	L	80	30
Fort Myers	40	95	M	78	25
Gainesville	30	95	M	78	30
Jacksonville	30	95	M	78	30
Key West	55	100	L	78	25
Miami	45	90	L	79	25
Orlando	35	90	M	78	30
Pensacola	25	95	L	78	30
Tallahassee	25	95	M	78	30
Tampa	35	95	M	78	30
GEORGIA					
Athens	10	95	M	76	35
Atlanta	10	95	M	76	35
Augusta	20	100	M	76	35
Brunswick	25	95	L	78	30
Columbus	20	100	M	76	35
Macon	20	95	M	78	35
Rome	10	95	M	76	35
Savannah	25	95	M	78	30
Way Cross	25	95	M	78	30
IDAHO					
Boise	-10	95	H	65	45
Idaho Falls	-15	90	H	65	45
Lewistown	-10	95	H	65	45
Pocatello	-15	90	H	65	45
Twin Falls	-15	95	H	65	40
ILLINOIS					
Aurora	-10	95	M	75	40
Bloomingtown	-10	95	M	76	40
Cairo	0	100	M	78	35
Champaign	-10	95	M	77	40
Chicago	-10	95	M	75	40
Danville	-10	95	M	77	40
Decatur	-10	95	M	77	40
Elgin	-15	95	M	78	40
Joliet	-10	95	M	76	40
Moline	-10	95	M	76	40
Peoria	-15	95	M	76	40
Rockford	-15	95	M	78	40
Rock Island	-10	95	M	76	40
Springfield	-10	95	M	77	40
Urbana	-10	95	M	77	40
INDIANA					
Elkhart	-10	95	M	75	40
Evansville	-5	95	M	78	40
Fort Wayne	-5	95	M	75	40
Indianapolis	-10	95	M	76	40
Lafayette	-10	95	M	76	40
South Bend	-10	95	M	75	40
Terre Haute	-5	95	M	78	40
IOWA					
Burlington	-10	95	M	78	40
Cedar Rapids	-15	95	M	78	40
Charles City	-20	95	M	75	45
Clinton	-15	95	M	78	40
Council Bluffs	-15	100	M	78	40
Davenport	-10	95	M	78	40
Des Moines	-15	95	M	78	40
Dubuque	-15	95	M	78	40
Fort Dodge	-15	95	M	78	40
Keokuk	-15	95	M	78	40
Marshalltown	-15	95	M	78	40
Sioux City	-15	95	M	78	40
Waterloo	-15	95	M	78	40
KANSAS					
Atchison	-10	100	M	76	40
Concordia	-10	95	M	78	40
Dodge City	-10	95	H	78	40
Iola	-5	100	M	75	40
Leavenworth	-10	100	M	76	40
Salina	-10	100	M	78	40
Topeka	-10	100	M	78	40
Wichita	-5	100	M	75	40

Table 15-2. Outside Design Conditions for the United States and Canada (continued from page 201).

State & City	Winter DB	Summer DB	Daily Range	Summer WB	Latitude Deg.
KENTUCKY					
Bowling Green	0	95	M	78	35
Frankfort	0	95	M	78	40
Hopkinsville	0	95	M	78	35
Lexington	0	95	M	78	40
Louisville	0	95	M	78	40
Owensboro	0	95	M	78	40
Shelbyville	0	95	M	78	40
LOUISIANA					
Alexandria	20	95	L	78	30
Baton Rouge	20	95	M	80	30
New Orleans	25	95	L	80	30
Shreveport	15	95	M	78	35
MAINE					
Augusta	-15	85	L	73	45
Bangor	-20	85	L	73	45
Bar Harbor	-10	85	L	73	45
Belfast	-10	85	L	73	45
Eastport	-10	85	L	70	45
Lewiston	-10	85	L	73	45
Millinocket	-15	85	M	73	45
Orono	-20	85	M	70	45
Portland	-10	85	M	73	45
Presque Isle	-20	85	L	73	45
Rumford	-15	85	L	73	45
MARYLAND					
Annapolis	10	90	M	78	40
Baltimore	10	90	M	78	40
Cambridge	10	90	L	78	40
Cumberland	0	90	M	75	40
Fredrick	5	90	M	78	40
Frostburg	-5	90	M	75	40
Salisbury	10	90	M	78	40
MASSACHUSETTS					
Amherst	-5	90	M	75	40
Boston	0	85	M	74	40
Fall River	0	85	L	75	40
Fitchburg	-5	90	M	75	45
Framingham	-5	85	L	75	40
Lawrence	-5	85	L	74	40
Lowell	-5	85	L	74	45
Nantucket	0	85	L	75	40
New Bedford	0	85	L	75	40
Pittsfield	-10	90	M	75	40
Plymouth	0	85	L	75	40
Springfield	-5	90	M	75	40
Worcester	-5	90	M	75	40
MICHIGAN					
Alpena	-10	90	M	75	45
Ann Arbor	-5	90	M	75	40
Big Rapids	-5	90	M	75	45
Cadillac	-10	90	M	75	45
Calumet	-20	80	M	73	45
Detroit	-5	90	M	75	40
Escanaba	-20	85	M	74	45
Flint	-10	90	M	75	45
Grand Haven	-5	90	M	75	45
Grand Rapids	-5	90	M	74	45
Houghton	-20	80	M	73	45
Kalamazoo	-5	90	M	75	40
Lansing	-10	90	M	75	45
Ludington	-5	90	M	75	45
Marquette	-15	80	M	73	45
Muskegon	-5	90	M	74	45
Port Huron	-10	90	M	75	45
Saginaw	-10	90	M	75	45
Sault Ste. Marie	-20	80	M	71	45
MINNESOTA					
Alexandria	-25	85	M	74	45
Duluth	-25	80	M	71	45
Minneapolis	-25	90	M	76	45
Moorhead	-30	95	M	75	45
St. Cloud	-25	90	M	76	45
St. Paul	-25	90	M	75	45

State & City	Winter DB	Summer DB	Daily Range	Summer WB	Latitude Deg.
MISSISSIPPI					
Biloxi	25	90	L	80	30
Columbus	10	95	M	78	35
Corinth	5	95	M	78	35
Hattiesburg	20	95	M	80	30
Jackson	15	95	M	78	30
Meridian	15	95	M	79	30
Natchez	15	95	L	78	30
Vicksburg	15	95	L	78	30
MISSOURI					
Columbia	−10	100	M	78	40
Hannibal	−10	95	M	77	40
Kansas City	−10	100	M	76	40
Kirksville	−10	95	M	78	40
St. Joseph	−10	100	M	76	40
St. Louis	−5	95	M	78	40
Springfield	−5	100	M	77	40
MONTANA					
Anaconda	−30	85	H	59	45
Billings	−30	90	H	66	45
Butte	−30	85	H	59	45
Great Falls	−40	90	H	63	50
Havre	−40	95	M	70	50
Helena	−40	90	H	63	45
Kalispell	−30	90	H	63	50
Miles City	−35	95	H	69	45
Missoula	−30	90	H	63	45
NEBRASKA					
Grand Island	−15	100	H	75	40
Hastings	−15	100	M	75	40
Lincoln	−15	95	M	78	40
Norfolk	−15	95	M	78	40
North Platte	−15	100	H	73	40
Omaha	−15	100	M	78	40
Valentine	−20	95	M	78	45
York	−15	95	M	78	40
NEVADA					
Elko	−10	95	H	63	40
Las Vegas	10	110	H	71	35
Reno	5	95	H	65	40
Tonopah	5	90	M	63	40
Winnemucca	−10	95	H	65	40
NEW HAMPSHIRE					
Berlin	−15	85	H	73	45
Claremont	−15	85	M	73	45
Concord	−10	85	H	73	45
Franklin	−15	85	M	73	45
Hanover	−15	85	M	73	45
Keene	−10	85	L	73	45
Manchester	−10	88	M	74	45
Nashua	−10	85	L	74	45
Portsmouth	−5	85	L	74	45
NEW JERSEY					
Asbury Park	5	90	L	78	40
Atlantic City	10	90	I	78	40
Bayonne	0	90	L	75	40
Belvidere	0	90	M	75	40
Bloomfield	0	90	L	75	40
Bridgeton	5	90	L	78	40
Camden	5	90	L	78	40
East Orange	0	90	L	75	40
Elizabeth	0	90	L	75	40
Jersey City	0	90	L	75	40
Newark	0	90	M	76	40
New Brunswick	5	90	L	75	40
Paterson	0	90	L	75	40
Phillipsburg	0	90	M	75	40
Trenton	0	90	L	78	40
NEW MEXICO					
Albuquerque	10	95	M	65	35
El Morro	0	85	H	65	35
Raton	−5	95	H	65	35
Roswell	5	100	H	71	35
Santa Fe	5	90	M	65	35
Tucumcari	5	95	H	70	35

Table 15-2. Outside Design Conditions for the United States and Canada (continued from page 204).

State & City	*Winter DB*	*Summer DB*	*Daily Range*	*Summer WB*	*Latitude Deg.*
NEW YORK					
Albany	-10	90	M	74	45
Auburn	-10	90	M	74	45
Binghampton	-5	90	M	72	40
Buffalo	-5	85	M	73	45
Canton	-20	85	M	73	45
Cortland	-10	90	M	74	45
Elmira	-5	90	M	73	40
Glens Falls	-15	90	M	73	45
Ithaca	-5	90	M	73	40
Jamestown	-5	90	M	74	40
Lake Placid	-15	90	M	73	45
New York	5	90	M	76	40
Niagara Falls	-5	85	M	73	45
Ogdensburg	-20	85	M	73	45
Onconta	-10	90	M	73	45
Oswego	-5	90	M	74	45
Port Jervis	0	90	L	75	40
Rochester	-5	90	M	74	45
Schenectady	-10	90	M	74	45
Syracuse	-10	90	M	74	45
Watertown	-15	85	M	73	45
NORTH CAROLINA					
Asheville	5	90	M	75	35
Charlotte	15	95	M	78	35
Greensboro	10	90	M	76	35
Hatteras	20	90	L	80	35
New Bern	20	95	L	78	35
Raleigh	15	95	M	78	35
Salisbury	10	90	M	78	35
Wilmington	20	90	M	81	35
Winston-Salem	10	90	M	76	35
NORTH DAKOTA					
Bismarck	-30	95	H	73	45
Devils Lake	-30	90	M	70	50
Dickinson	-30	95	H	70	45
Fargo	-30	95	H	75	45
Grand Forks	-30	90	M	72	50
Jamestown	-30	95	M	73	45
Minot	-35	90	M	71	50
Pembina	-35	90	M	73	50
Williston	-35	90	M	73	50
OHIO					
Akron	-5	90	M	75	40
Cincinnati	-5	95	M	78	40
Cleveland	-5	90	M	75	40
Columbus	-5	90	M	76	40
Dayton	-5	90	M	76	40
Lima	-5	90	M	75	40
Marion	-5	90	M	75	40
Sandusky	-5	90	M	75	40
Toledo	-5	90	M	75	40
Warren	-5	90	M	75	40
Youngstown	-5	90	M	75	40
OKLAHOMA					
Ardmore	5	100	M	78	35
Bartlesville	-5	100	M	77	35
Guthrie	0	100	M	77	35
Muskogee	0	95	M	79	35
Oklahoma City	0	100	M	77	35
Tulsa	0	100	M	77	35
Waynoka	-5	105	M	75	35
OREGON					
Arlington	5	95	M	68	45
Baker	-15	90	M	66	45
Eugene	15	90	H	68	45
Medford	20	95	H	68	40
Pendleton	-10	90	H	66	45
Portland	10	85	M	68	45
Roseburg	20	90	H	66	45
Salem	15	90	H	68	45
Wamic	0	90	H	66	45

State & City	Winter DB	Summer DB	Daily Range	Summer WB	Latitude Deg.
PENNSYLVANIA					
Altoona	−5	90	M	75	40
Bethlehem	0	90	M	75	40
Coatesville	5	90	M	75	40
Erie	−5	85	M	74	40
Harrisburg	5	90	M	75	40
New Castle	−5	90	M	75	40
Oil City	−5	90	M	75	40
Philadelphia	5	90	M	78	40
Pittsburgh	−5	90	M	75	40
Reading	5	90	M	75	40
Scranton	0	90	M	75	40
Warren	−5	90	M	75	40
Williamsport	−5	90	M	74	40
York	5	90	M	75	40
RHODE ISLAND					
Block Island	5	85	L	75	40
Bristol	0	90	L	75	40
Kingston	0	85	L	75	40
Pawtucket	0	90	M	75	40
Providence	0	90	M	75	40
SOUTH CAROLINA					
Charleston	20	90	L	80	35
Columbia	20	95	M	78	35
Florence	20	95	M	79	35
Greenville	10	95	M	75	35
Spartanburg	10	95	M	78	35
SOUTH DAKOTA					
Aberdeen	−25	95	M	75	45
Huron	−20	100	H	75	45
Pierre	−20	95	M	73	45
Rapid City	−20	95	H	70	45
Sioux Falls	−20	95	H	75	45
Watertown	−25	95	M	73	45
TENNESSEE					
Chattanooga	10	95	M	76	35
Jackson	5	95	M	78	35
Johnson City	0	95	M	78	35
Knoxville	5	95	M	78	35
Memphis	5	95	M	78	35
Nashville	5	95	M	78	35
TEXAS					
Abilene	5	95	M	74	30
Amarillo	0	95	H	72	35
Austin	15	100	M	78	30
Brownsville	30	95	M	80	25
Corpus Christi	25	95	M	80	30
Dallas	10	100	M	78	35
Del Rio	20	100	H	78	30
El Paso	20	100	M	69	30
Fort Worth	10	100	M	78	35
Galveston	25	95	L	80	30
Houston	20	95	M	80	30
Palestine	10	100	M	78	30
Port Arthur	20	95	M	80	30
San Antonio	20	100	M	78	30
Waco	10	100	M	78	30
UTAH					
Logan	−10	95	H	65	40
Milford	-5	95	H	66	40
Ogden	−5	90	H	65	40
Salt Lake City	0	95	H	65	40
VERMONT					
Bennington	−10	90	M	73	45
Burlington	−15	90	M	73	45
Montpelier	−20	90	M	73	45
Newport	−20	85	M	73	45
Northfield	−20	90	M	73	45
Rutland	−15	90	M	73	45

Table 15-2. Outside Design Conditions for the United States and Canada (continued from page 206).

State & City	*Winter DB*	*Summer DB*	*Daily Range*	*Summer WB*	*Latitude Deg.*
VIRGINIA					
Cape Henry	15	90	L	78	35
Charlottesville	10	90	M	78	40
Danville	10	90	M	78	35
Lynchburg	10	90	M	76	35
Norfolk	15	90	L	78	35
Petersburg	10	90	M	78	35
Richmond	10	90	M	78	40
Roanoke	5	90	M	76	35
Wytheville	5	90	M	76	35
WASHINGTON					
Aberdeen	20	85	L	64	45
Bellingham	10	80	L	65	50
Everett	15	80	L	65	50
North Head	20	80	L	65	50
Olympia	15	80	M	64	45
Seattle	15	80	M	65	50
Spokane	-15	80	H	65	50
Tacoma	15	80	M	64	45
Tatoosh Island	20	80	L	65	50
Walla Walla	-10	90	H	65	45
Wenatchee	-10	90	M	65	50
Yakima	-5	90	H	67	45
WEST VIRGINIA					
Bluefield	0	95	M	75	35
Charleston	0	90	M	75	40
Elkins	-5	90	M	73	40
Fairmont	0	90	M	75	40
Huntington	0	90	M	76	40
Martinsburg	0	90	M	75	40
Parkersburg	0	90	M	75	40
Wheeling	-5	90	M	75	40
WISCONSIN					
Ashland	-25	80	M	71	45
Beloit	-15	95	M	78	45
Eau Claire	-20	90	M	75	45
Green Bay	-20	90	M	73	45
La Crosse	-20	95	M	75	45
Madison	-20	90	M	75	45
Milwaukee	-15	90	M	75	45
Oshkosh	-20	90	M	75	45
Sheboygan	-20	90	M	75	45
WYOMING					
Casper	-125	90	H	62	45
Cheyenne	-20	90	H	62	40
Lander	-30	90	H	65	45
Sheridan	-30	90	H	65	45
Yellowstone Park	-35	85	H	62	45

Province & City	*Winter DB*	*Summer DB*	*Daily Range*	*Summer WB*	*Latitude Deg.*
ALBERTA					
Banff	-30	–	H	–	50
Camrose	-35	–	H	–	55
Calgary	-30	90	H	66	50
Cardston	-30	–	H	–	50
Edmonton	-35	90	H	68	55
Grande Prairie	-40	–	H	–	55
Hanna	-35	–	H	–	50
Jasper	-30	–	H	–	55
Lethbridge	-30	–	H	–	50
Lloydminster	-40	–	H	–	55
McMurray	-40	–	H	–	55
Medicine Hat	-35	90	H	65	50
Red Deer	-35	–	H	–	50
Taber	-35	–	H	–	50
Wetaskiwin	-35	–	H	–	55
BRITISH COLUMBIA					
Chilliwack	5	–	M	–	50
Courtenay	10	–	M	–	50
Dawson Creek	-40	–	H	–	55
Estevan Point	15	–	M	–	55
Fort Nelson	-40	–	H	–	60
Hope	0	–	M	–	50
Kamloops	-20	–	H	–	50

Province & City	*Winter DB*	*Summer DB*	*Daily Range*	*Summer WB*	*Latitude Deg.*
Kimberly	-25	-	H	-	50
Lytton	-5	-	H	-	50
Nanaimo	10	-	M	-	50
Nelson	-10	-	H	-	50
Penticton	-5	-	H	-	50
Port Alberni	10	-	M	-	50
Prince George	-30	-	H	-	55
Prince Rupert	10	-	L	-	55
Princeton	-15	-	H	-	50
Revelstoke	-25	-	H	-	50
Trail	-10	-	H	-	50
Vancouver	10	80	L	67	50
Vernon	-15	-	H	-	50
Victoria	15	-	L	-	50
Westview	10	-	M	-	50
LABRADOR					
Goose Bay	-25	-	L	-	55
MANITOBA					
Boissevain	-35	-	H	-	50
Brandon	-30	-	H	-	50
Churchill	-40	-	L	-	60
Dauphin	-35	-	M	-	50
Flin Flon	-40	-	M	-	55
Minnedosa	-35	-	H	-	50
Neepawa	-35	-	H	-	50
La Prairie	-30	-	M	-	50
Swan River	-35	-	M	-	55
The Pas	-40	-	M	-	55
Winnipeg	-30	90	M	71	50
NEW BRUNSWICK					
Bathurst	-10	-	L	-	45
Campbellton	-10	-	L	-	45
Chatham	-10	-	M	-	45
Edmuntson	-15	-	M	-	45
Fredericton	-5	90	L	75	45
Moncton	-10	-	M	-	45
Saint John	-5	80	L	67	45
Woodstock	-15	-	M	-	45
NEWFOUNDLAND					
Corner Brook	0	-	L	-	50
Gander	-5	-	L	-	50
Grand Falls	-5	-	M	-	50
St. John's	0	-	L	-	50
NORTHWEST TERRITORIES					
Aklavik	-45	-	L	-	70
Fort Norman	-40	-	M	-	65
Frobisher	-50	-	L	-	-
Resolute	-40	-	L	-	-
Yellowknife	-50	-	L	-	60
NOVA SCOTIA					
Bridgewater	0	-	L		45
Dartmouth	0	-	L	-	45
NOVA SCOTIA					
*Halifax C	5	80	L	75	45
*Halifax A	0	80	L	75	45
Kentville	0	-	L		45
New Glasgow	0	-	L	-	45
Spring Hill	-5	-	M	-	45
Sydney	0	85	L	67	45
Truro	0	-	L	-	45
Yarmouth	5	-	L	-	45
ONTARIO					
Bancroft	-20	-	M	-	45
Barrie	-5	-	M	-	45
Belleville	-10	-	M	-	45
Brampton	-5	-	M	-	45
Brantford	-5	-	M	-	45
Brockville	-15	-	M	-	45
Chatham	0	-	M	-	45
Cobourg	-10	-	M	-	45
Collingwood	0	-	M	-	45

Table 15-2. Outside Design Conditions for the United States and Canada (continued from page 207).

Cornwall	−15	–	M	–	45
Ear Falls	−35	–	M	–	50
Fort Frances	−30	–	M	–	50
Fort William	−25	85	M	70	50
Galt	−5	–	M	–	45
Geraldton	−35	–	M	–	50
Goderich	0	–	M	–	45
Guelph	−5	–	M	–	45
Hamilton	0	–	M	–	45
Haileybury	−25	–	M	–	50
Hanover	−5	–	M	–	45
Huntsville	−15	–	M	–	45
Kapuskasing	−30	–	M	–	50
Kenora	−35	–	M	–	50
Kingston	−10	–	M	–	45
Kirkland Lake	−25	–	M	–	50
Kitchener	−5	–	M	–	45
Lindsay	−15	–	M	–	45
London	0	–	M	–	45
Moonsonee	−35	–	M	–	50

Table 15-2, use the design temperatures of listed cities that most closely approximate your local conditions.

The inside dry-bulb temperature for heating in colder months is entirely up to you. Most people prefer a temperature of from 70 to 75°F. The temperature difference will be the difference between the inside design temperature (say 70°F) and the outside design conditions for your locality.

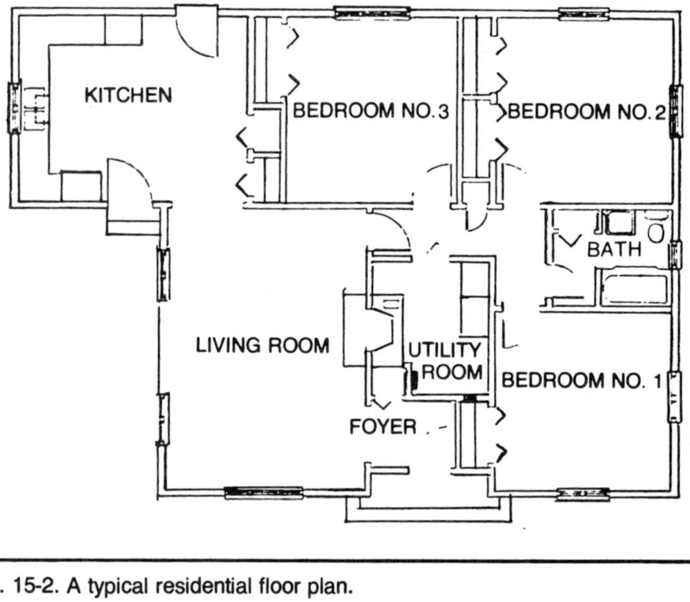

Fig. 15-2. A typical residential floor plan.

In order to demonstrate exactly how to make heat loss calculations for a given home, let's use the floor plan of the residence in Fig. 15-2 as an example. Assume that this home is located near Washington D.C., and use design conditions for this area. With the inside and outside design conditions and the drawing of the house floor plan, you are ready to do a complete room by room heat loss calculation for the house. Use a separate form for each room in the house.

Living Room

Begin by filling in Part 1 of the form for the living area (living room) (Table 15-3). This information will include the following:

- Ceiling below ventilated attic—2 inches of insulation.
- Floor with crawl space—2 inches of insulation.
- Windows—single pane with drapes.
- Walls and insulation—plasterboard with 3½ inches of insulation.

Part 2 of the form covers heat loss as a result of infiltration, or air leaks through cracks in walls and around windows and doors. The example living room is 13.75 feet wide by 19.25 feet. It has an 8-foot ceiling. When these dimensions are inserted in the proper blanks, you can complete the formula as follows:

$$\frac{13.75 \times 19.25 \times 8}{60} = 35.29$$

Part 3 of the form is outside walls: one of the living room walls is 19.25 feet in length, while the remaining outside wall is 13.75 feet

Table 15-3. Completed Heat Loss Form for the Living Room of the Residence in Fig. 15-2.

NAME OF PROJECT OR AREA LIVING ROOM
DESIGN CONDITIONS: Outside Dry-Bulb Temperature 0°
Inside Dry-Bulb Temperature 70° Temperature Difference 70°F.

Part 1: Ceiling Type and Insulation Thickness 2 inches
Floor Type and Insulation Crawl Space 2 inches of insulation
Type of Windows Single Pane
Type of Walls and Insulation 3½ inches of insulation
Part 2: Infiltration
area width × area length × ceiling heat
(13.75) width × (19.25) length × (8) ceiling height
60 Heat Loss
= 35.29 × 1.08 × 70 T.D. = 2667.70

Part 3: Heat Loss

Windows:	49	square feet × 0.55 × 70 T.D.=	1886.5
Walls:	215	square feet × 0.13 × 70 T.D.=	1956.5
Roof:	264.68	square feet × 0.20 × 70 T.D.=	3705.1
Floor:	264.68	square feet × 0.09 × 70 T.D.=	1668.80

TOTAL HEAT LOSS IN Btu/h
11,884.6
Divide by 3.4 to obtain heat loss in watts 3,495.5

Table 15-4. Completed Heat Loss Form for Bedroom Number 1.

NAME OF PROJECT OR AREA Bedroom Number 1
DESIGN CONDITIONS: Outside Dry-Bulb Temperature 0°
Inside Dry-Bulb Temperature 70° Temperature Difference 70°F

Part 1: Ceiling Type and Insulation Thickness 2 inches
Floor Type and Insulation 2 inches
Type of Windows Single Pane
Type of Walls and Insulation 3½ inches

Part 2: Infiltration

$$\frac{\text{area width} \times \text{area length} \times \text{ceiling height}}{\dfrac{(12)\ \text{width} \times (12.5)\ \text{length} \times (8)\ \text{ceiling height}}{60}}$$

Heat Loss
= 20 × 1.08 × 70 T.D. = 1512.0

Part 3: Heat Loss

Windows:	28	square feet × 0.55 ×	70	T.D.=	1078.0
Walls:	164	square feet × 0.13 ×	70	T.D.=	1492.4
Roof:	144	square feet × 0.20 ×	70	T.D.=	2016.0
Floor:	144	square feet × 0.09 ×	70	T.D.=	907.2

TOTAL HEAT LOSS IN Btu/h 7005.6
Divide by 3.4 to obtain heat loss in watts 2060.47

in length. Since you know the ceiling height is 8 feet, the gross outside wall area may be found by:

$$19.25 \text{ feet} \times 8 \text{ feet} = 264 \text{ square feet}$$

The calculation form calls for the net wall area (gross wall area-window area), so you must subtract 49 square feet from 264 square feet to obtain the answer—215 square feet. Again, this figure is entered in the appropriate space on the form.

The remaining two spaces are for the areas of the floor and ceiling. Since these two areas will have the same dimensions, you can complete the calculation for both in one step:

$$19.25 \times 13.75 = 264.68 \text{ square feet (of floor and ceiling)}$$

This completes all the data that is necessary for the calculation of the heat loss which must be taken from the drawings. The remaining factors are printed on the form.

Now begin with the first figure (infiltration). Multiply by the infiltration factor (1.08) and the temperature differential (7°), and enter the product in the appropriate space. Continue down the list until heat loss is calculated for all areas listed. Summing the subtotals, you find the room heat loss is 11,884.6 Btu/h.

This means that 11,884.6 Btu/h will be required to keep the inside temperature at 70°F if the outside temperature is 0°F. The complete heat loss form for the living room is Table 15-3. Briefly review this entire procedure before continuing on to the next area.

If you were adding a woodburning stove only to the living room—like a woodburning insert for the fireplace—and heating the remaining areas with the existing heating system, you could stop now. This method of calculating heat loss is good for one small area or an entire home.

Bedrooms and Bathroom

Let's assume that you are installing a woodburning furnace to heat the entire house—making it necessary to find the heat loss for the remaining areas. The kitchen/dining area, bedrooms, and bathroom calculations are performed exactly as for the living area, using different dimensions. The results of the calculations are shown in Tables 15-4 through 15-7.

Utility Room

The remaining area—the utility room—is calculated in a slightly different way since there are no outside walls or windows (all surrounding areas will be heated). Therefore, these steps may be omitted on the form. The calculated results for this area are shown in Table 15-8.

OTHER CALCULATIONS

Another quick way to determine the required heat in Btu for a particular area in which you're installing a woodburning heater is to check existing heating equipment of a different type. For example, if

Table 15-5. Completed Heat Loss Form for Bedroom Number 2.

NAME OF PROJECT OR AREA Bedroom Number 2
DESIGN CONDITIONS: Outside Dry-Bulb Temperature 0°
Inside Dry-Bulb Temperature 70° Temperature Difference 70°

Part 1: Ceiling Type and Insulation Thickness 2 inches
Floor Type and Insulation 2 inches
Type of Windows Single Pane
Type of Walls and Insulation 3½ inches

Part 2: Infiltration

area width × area length × ceiling height

(12.25) width × (15.5) length × (8) ceiling height / 60 = 20.41 × 1.08 × 70 T.D.= 1542.8 Heat Loss

Part 3: Heat Loss

Windows:	28	square feet × 0.55 ×	70	T.D.=	1078.0
Walls:	172	square feet × 0.13 ×	70	T.D.=	1565.2
Roof:	156	square feet × 0.20 ×	70	T.D.=	2184.0
Floor:	156	square feet × 0.90 ×	70	T.D.=	982.8

TOTAL HEAT LOSS IN Btu/h 7352.8
Divide by 3.4 to obtain heat loss in watts 2162.58

Table 15-6. Completed Heat Loss for Bedroom Number 3.

NAME OF PROJECT OR AREA Bedroom Number 3
DESIGN CONDITIONS: Outside Dry-Bulb Temperature 0°
Inside Dry-Bulb Temperature 70° Temperature Difference 70°F

Part 1: Ceiling Type and Insulation Thickness 2 inches
Floor Type and Insulation 2 inches
Type of Windows Single Pane
Type of Walls and Insulation 3½ inches

Part 2: Infiltration
area width × area length × ceiling height

$$\frac{(12)\text{ width} \times (12.5)\text{ length} \times (8)\text{ ceiling height}}{60}$$

Heat Loss

= 20 × 1.08 × 70 T.D. = 1512.0

Part 3: Heat Loss

Windows:	21	square feet × 0.55 ×	70 T.D.=	808.5
Walls:	75	square feet × 0.13 ×	70 T.D.=	682.5
Roof:	156	square feet × 0.20 ×	70 T.D.=	2184.0
Floor:	156	square feet × 0.90 ×	70 T.D.=	982.8

TOTAL HEAT LOSS IN Btu / h 6169.8

Divide by 3.4 to obtain heat loss in watts 1814.69

the living room in the example is presently heated with electric baseboard heaters, check the nameplate of the electric heating units for their ratings. Let's assume that the living room now has electric heaters totaling 3500 watts, and you want to know the size wood heater required to heat the same area. Merely multiply the total watts by the factor 3.4, or 3500 × 3.4 = 11,900 Btu. This is close to the original calculation of 11,884.6.

Table 15-7. Completed Heat Loss Form for Bathroom.

NAME OF PROJECT OR AREA Bathroom
DESIGN CONDITIONS: Outside Dry-Bulb Temperature 0°
Inside Dry-Bulb Temperature 70° Temperature Difference 70°F
Part 1: Ceiling Type and Insulation Thickness 2 inches
Floor Type and Insulation 2 inches
Type of Windows Single Pane
Type of Walls and Insulation 3½ inches

Part 2: Infiltration
area width × area length × ( ) ceiling height

$$\frac{(6.5\text{ width}) \times (8.25)\text{ length} \times (8)\text{ ceiling height}}{60}$$

Heat Loss

= 7.14 × 1.08 × 70 T.D. = 539.7

Part 3: Heat Loss

Windows:	4	square feet × 0.55 ×	70 T.D.=	154.0
Walls:	48	square feet × 0.13 ×	70 T.D.=	436.8
Roof:	52	square feet × 0.20 ×	70 T.D.=	728.0
Floor:	52	square feet × 0.90 ×	70 T.D.=	327.6

TOTAL HEAT LOSS IN Btu / h 2186.1
Divide by 3.4 to obtain heat loss in watts 642.97

Table 15-8. Completed Heat Loss Form for Utility Room

NAME OF PROJECT OR AREA Utility Room
DESIGN CONDITIONS: Outside Dry-Bulb Temperature 0°
Inside Dry-Bulb Temperature 70° Temperature Difference 70°F

Part 1: Ceiling Type and Insulation Thickness 2 inches
Floor Type and Insulation 2 inches
Type of Windows Single Pane
Type of Walls and Insulation 3½ inches
Part 2: Infiltration
area width × area length × ceiling weight

$$\frac{(5.5)\text{ width} \times (9)\text{ length} \times (8)\text{ ceiling height}}{60}$$ Heat Loss

= 6.6 × 1.08 × ___ T.D. = 498.4

Part 3: Heat Loss

Windows:	___	square feet × 0.55 ×	___ =	___
Walls:	___	square feet × 0.13 ×	___ =	___
Roof:	49.5	square feet × 0.20 ×	70 T.D.=	693.0
Floor:	49.5	square feet × 0.90 ×	70 T.D.=	311.5

TOTAL HEAT LOS IN Btu / h 1502.9

Divide by 3.4 to obtain heat loss in watts 442.02

Some homeowners are now using hot water baseboard heat. The water is heated by coils through a woodburning unit. A simple rule of thumb equation or method to use in determining the amount of hot water radiating surface required for a given area is to allow 1 square foot of radiating surface for every 2 square feet of glass surface, one for every 20 feet of outside walls, and one for every 200 cubic feet of room contents. In the living room example, you would need the following square feet of radiation:

$$\frac{49}{2}\text{ square feet of glass} = 24.5 \text{ square feet radiation}$$

$$\frac{215}{20}\text{ square feet walls} = 10.75 \text{ square feet radiation}$$

$$\frac{2117}{200}\text{ cubic feet} = 10.58 \text{ square feet radiation}$$

Total square feet of radiation = 45.88

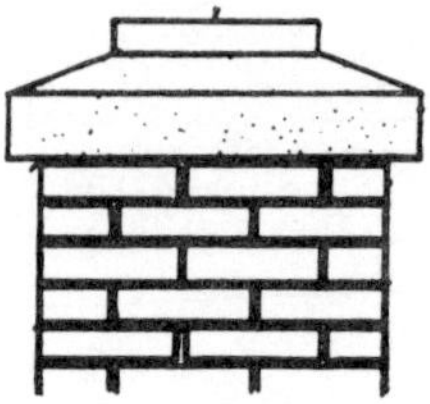

Glossary

actual evaporation—The total quantity of water (in pounds) evaporated from the temperature of the feed water to steam at 212°F.

air cleaner—Device used for removal of airborne impurities.

air-dry—Wood containing 20 percent moisture or less.

airtight stove—One with all seams welded or otherwise sealed and having tight fitting doors. Air can enter the stove only at the draft control, giving full control over the burning rate.

ambient temperature—Temperature of fluid (usually air) which surrounds an object on all sides.

atmospheric pressure—The pressure exerted by the atmosphere may be established by the simple experiment of taking a glass tube, with an area of 1 square inch and approximately 30 inches high, with one end closed. Fill it with mercury and invert the tube in mercury. At sea level the mercury will stand in the column 29.9 inches high, or practically 30 inches. As 1 cubic foot of mercury weighs 850 pounds, 1 cubic inch would weigh 0.49 pounds. As the mercury in the tube under the pressure of the atmosphere stands 30 inches high, and the area is 1 square inch, there will be 30 cubic inches of mercury in the tube, which will weigh 30 × 0.49, or 14.7 pounds. As this column of mercury is entirely sustained by the pressure of the atmosphere, it may be stated that the normal pressure at sea level is 14.7 pounds per square inch. This pressure varies with the altitude and under different conditions of the barometer.

barometer—Instrument for measuring atmospheric pressure. It may be calibrated in pounds per square inch or in inches of mercury in column.

bellows—Corrugated cylindrical container which moves as pressure changes or provides a seal during movement of parts.

bimetal strip—Temperature regulating or indicating device which works on the principle that two dissimilar metals with unequal expansion rates, welded together, will bend as temperatures change.

boiler—Closed container in which a liquid may be heated and vaporized.

boiler heating surface, direct—That surface which receives the radiant heat of the fire, or that surface on which the fire shines.

boiler heating surface, indirect flue—That surface in a boiler on which the fire does not shine, but through which the constantly cooling gases pass to the smokestack. The value of indirect, or flue surface, is extremely variable because the escaping gases are constantly cooling; the rate of transmission becomes less and less as the gases approach the smoke outlet.

boiling temperature—Temperature at which a fluid changes from a liquid to a gas.

Btu (British thermal unit)—The amount of heat required to raise the temperature of 1 pound of water 1°F.

burner—Device in which combustion of fuel takes place.

butane—Liquid hydrocarbon (C_4H_{10}) commonly used as fuel for heating purposes.

bypass—Passage at one side of or around regular passage.

calorie—Heat required to raise the temperature of 1 gram of water 1°C.

carbon dioxide—Compound of carbon and oxygen (CO_2) which is sometimes used as a refrigerant.

carbon filter—Air filter using activated carbon as air cleansing agent.

Celsius—German language word for centigrade, the metric system temperature scale.

centigrade scale—Temperature scale used in metric system. The freezing point of water is 0; the boiling point is 100.

charcoal—The remainder of wood after all water and volatile gases have been heated and driven off. It consists mainly of carbon and gives off intense heat.

check value—A device which permits fluid flow only in one direction.

chimney—An upright structure for conducting smoke and gases from the house. It may be heavy masonry or a prefabricated pipe consisting of two or more layers of steel with insulation between.

chimney effect—Tendency of air or gas to rise when heated.

circulator—Common term for one type of airtight stove with a jacket or cabinet around the basic stove/firebox system; it is usually offered with an optional blower system.

code installation—A heating installation which conforms to the local code and/or the national code for safe and efficient installations.

coefficient of conductivity—The measure of the relative rate at which different materials conduct heat. Copper is a good conductor of heat and has a high coefficient of conductivity.

coefficient of heat emission—This term is usually applied to the heat emitted, or given off, by 1 square foot of radiation per hour for 1°

temperature difference between the steam or water in the radiator and the surrounding air.

coefficient of transmission—The quantity of heat, expressed in terms of Btu, which will pass through 1 square foot of surface in 1 hour for 1° temperature difference.

cold—Cold is the absence of heat, a temperature considerably below normal.

combustion chamber—That portion of the firebox between the surface of the fuel bed and the crown sheet of the heater.

combustion, rate of—The rate of combustion is a term usually applied to the quantity of coal burned per square foot of grate per hour. The term may also be used to designate the quantity of fuel burned by the boiler per hour.

comfort chart—Chart used in air conditioning to show the dry-bulb temperature and humidity for human comfort conditions.

comfort zone—Area on *psychrometric* chart which shows conditions of temperature, humidity, and sometimes air movement in which most people are comfortable.

condensate—The liquid resulting when water vapor contacts a cool surface.

condensing power (of radiation)—The condensing power of radiation is the quantity of steam (in pounds) which a radiator will condense per square foot per hour.

conduction—The transfer of heat between two bodies or parts of a body which touch each other. Internal conduction takes place between the parts of one continuous body, and external conduction occurs through the surface of contact of a pair of distinct bodies.

control—Automatic or manual device used to stop, start, and/or regulate the flow of a gas, liquid, and/or electricity.

control, temperature—A thermostatic device which automatically stops and starts a motor whose operation is based on temperature changes.

convection—Heat transfer by gravity movement of warmed room air to other parts of the house. It is one method by which a stove provides heat.

convection, forced—Transfer of heat resulting from forced movement of a liquid or gas by means of a fan or pump.

convection, natural—Circulation of a gas or liquid due to difference in density resulting from temperature differences.

cord—A stack of wood 4 feet wide × 4 feet high × 8 feet long containing a gross volume of 128 cubic feet, commonly referred to as a "full cord."

creosote—The black, tarlike substance deposited on the insides of stoves, smoke pipes, and chimneys. Its buildup in flue pipes and chimneys creates a fire hazard.

critical temperature—Temperature at which vapor and liquid have the same properties.

cycle—Series of events which have the tendency to repeat the events in the same order.

damper—A plate located inside the stovepipe collar, stovepipe, or at the base of a fireplace chimney to adjust the strength of the chimney draft.

degree-day—Unit that represents one degree of difference from a given point in the average outdoor temperature of one day. It is often used in estimating fuel requirements for a building. Degree-days are based on average temperature over one 24-hour period.

density—Closeness of texture or consistency.

draft control—A device used to change the amount of combustion air admitted to the stove firebox. It is also called a draft regulator.

draft gauge—Instrument used to measure air movement.

draft indicator—An instrument used to indicate or measure chimney draft or combustion gas movement. Draft is measured in units of .1 inch of water column.

dry-bulb—An instrument with a sensitive element which measures ambient (moving) air temperature.

dry-bulb temperature—Air temperature as indicated by an ordinary thermometer.

effective temperature—Overall effect on a human of air temperature, humidity, and air movement.

efficiency—Ratio of useful heat extracted to the total possible heat contained in the fuel.

equivalent evaporation—The total quantity of water (in pounds) evaporated from a temperature of 212° to steam at 212°F.

evaporation—The act of resolving, or the state of being resolved, into vapor. The conversion of boiling water by heat into vapor or steam.

evaporative power (of fuel)—The quantity of water (in pounds) which 1 pound of fuel will evaporate in burning. Theoretical evaporation is the quantity of water evaporated by 1 pound of fuel burning to perfect combustion.

face cord—A stack of wood 4 feet high and 8 feet long of wood pieces cut 12, 16, or 24 inches long; it is often confused with and sold as a "cord" or "full cord."

fan—A radial or axial flow device used for moving or producing artificial currents of air.

firebricks—Bricks used to line fireboxes in stoves and furnaces to extend their life. They are made of fireclay and can withstand high temperatures.

fire stop spacer—A noncombustible spacer used to provide clearance between chimney pipe and combustible walls or ceilings.

flash point—Temperature at which an oil will give off sufficient vapor to support a flash flame, but will not support continuous combustion.

flue—Gas or air passage which usually depends on natural convection to cause the combustion gases to flow through it. Forced convection may sometimes be used.

flue collar—A collar around the flue vent on the stove which fits into or over the flue pipe.
flue liner—Tiles (made of fireclay) used to line masonry chimneys.
freezing—Change of state from liquid to solid.
freezing point—The temperature at which a liquid will solidify upon removal of heat. The freezing temperature for water is 32°F at atmospheric pressure.

gas—Vapor phase or state of a substance.
gas noncondensable—A gas which will not form into a liquid under pressure-temperature conditions.
green wood—Wood that is still standing, or freshly cut, with the highest moisture content. It may be distinguished from air-dry wood by the lack of radial cracks.

hardwood—Tough, heavy timber with a compact texture. Generally, it is the wood of a broad-leafed tree.
heat—Added energy that causes substances to rise in temperature. It is the energy associated with the random motion of molecules.
heat exchanger—Device used to transfer heat from a warm or hot surface to a cold or cooler surface. Evaporators and condensers are heat exchangers.
heat load—Amount of heat, measured in Btu, which is removed during a period of 24 hours.
heat transmission—When the temperatures on opposite sides of any surface are unequal, the heat will flow through the material from the warmer to the cooler side. This is called heat transmission. Heat may be transferred by radiation, conduction, convection, or a combination of these three methods.
heating control—Device which controls temperature of the heat transfer unit which releases heat.
humidity—Moisture or dampness. Relative humidity is the ratio of the quantity of vapor present in the air to the greatest amount possible at a given temperature.
hydrometer—Floating instrument used to measure specific gravity of a liquid. Specific gravity is the ratio of weight of any volume of a substance to the weight of equal volume of a substance used as a standard.

ignition temperature—The temperature to which a substance must be raised in the presence of oxygen to cause the two to unite by combustion. It is rather indefinite and varies for different types of fuel. The exact temperature can only be determined by direct experimentation with the particular variety of fuel under test.
insulated pipe—Multiwalled chimney pipe used in wood stove installations to prevent loss of heat and maintain proper flue temperatures for sufficient draft.

insulation, thermal—Substance used to retard or slow the flow of heat through a wall or partition.

Kelvin scale (K)—Thermometer scale on which the unit of measurement equals the centigrade degree and according to which absolute zero is 0°, the equivalent of −273.16°C. Water freezes at 273.16° and boils at 373.16°.

latent heat—Heat energy absorbed in the process of changing the form of a substance (melting, vaporization, fusion) without a change in temperature or pressure.

leak detector—Device or instrument such as a *halide* torch or an electronic sniffer used to detect leaks.

LP fuel—Liquefied petroleum which is used as a fuel gas.

mass—A quantity of matter usually of considerable size.

non-code installation—A functional heating system installed where there are no local, state, or national codes in force.

potential heat—Maximum amount of heat available from any given species of wood. Potential heat figures for different woods (in Btu) are determined by burning the wood at zero moisture content under laboratory conditions.

radiant heating—Heating system in which warm or hot surfaces are used to radiate heat into the space to be conditioned.

radiation—Heat transfer from a stove or other hot surface directly to another surface such as a wall or furniture without affecting the temperature between them.

refractory—An asbestos substance mixed with concrete and applied to the bottom of a stove firebox as added protection against burnout.

relative humidity—Ratio of amount of water vapor present in air to the greatest amount possible at the same temperature.

seasoned wood—Wood that has been exposed to the air to the point that its moisture content remains constant with that of the surrounding atmosphere; this wood is more suitable for burning.

secondary air inlets—Air inlets on a stove designed to provide the necessary oxygen for secondary combustion to take place.

secondary combustion—Burning of flue gases inside a stove. Both oxygen and high temperatures must be present for this to occur.

sensible heat—Heat which causes a change in temperature of a substance.

sensor—A material or device which goes through a physical change or an electronic change as the conditions change.

single wall stovepipe—Metal pipe used to connect the stove to the outside chimney. Another term is flue pipe or chimney connector pipe.

smoke detector—Electronic device used to detect smoke in the air in unwanted areas to indicate a fire. Most detectors also have a built-in heat sensor that will sound an alarm when the ambient temperature reaches a predetermined level. Manufacturer's recommendations should be followed when installing these devices, as they are rated by the cubic feet of area they are to protect.

smoke test—Test made to determine the completeness of combustion.

softwood—Usually wood from conifers. Softwoods often are more resinous, giving them a slight edge in fuel value per pound of dry wood. Softwoods generally produce more creosote when burned, so they are not as desirable as most hardwoods for fuel.

specific gravity—Density of a substance divided by the density of water, where water is given the value of 1.0.

specific heat—Ratio of the quantity of heat required to raise the temperature of a body 1° to that required to raise the temperature of an equal mass of water 1°.

specific volume—Volume per unit mass of a substance.

stovepipe—Sometimes called smoke pipe. A single wall pipe venting the stove into the chimney.

steam—Water in vapor state.

steam heating—Heating system in which steam from a boiler is conducted to radiators in the space to be heated.

stoker—Machine used to supply a furnace with coal.

temperature—Degree of hotness or coldness as measured by a thermometer—the measurement of speed of motion of molecules.

temperature humidity index—Actual temperature and humidity of a sample of air compared to air at standard conditions.

therm—Quantity of heat equivalent to 100,000 Btu.

thermal mass—Material or mass of sufficient size and density to have heat storage capabilities. Thick-bodied stoves and massive hearths have sufficient thermal to store heat.

timer-thermostat—Thermostat control which includes a clock mechanism. The unit automatically controls room temperature and changes it according to the time of day.

Manufacturers

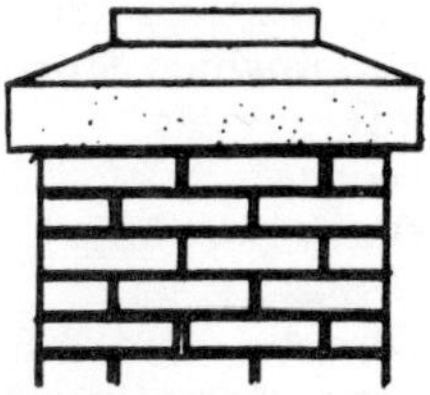

Abundant Life Farm, Inc.
P.O. Box W188
Lochmere, NH 03252
(Comforter multifuel stove)
Alaska Midwest, Inc.
520 N. Clark St.
River Falls, WI 92065
Albright Welding Co.
Rt. 15, Box 108
Jeffersonville, VT 05464
AllNighter Stove Works, Inc.
35 Nutmeg Lane
Glastonbury, CT 06033
Alternage Energy Sources, Inc.
752 Duvall
Salina, KS 67401
American Stovalator
Rt. 7
Arlington, VT 05250
Amerind Mackissic, Inc.
Box 111
Parker Ford, PA 19457
Amherst Welding, Inc.
330 Harkness Rd.
Amherst, MA 01002
Amity Woodstoves
P.O. Box 11
Athol, PA 19502
Antique Stove Works, Inc.
2505 Havana
Spokane, WA 99203
Aquappliances, Inc.
135-H Sunshine Lane
San Marcos, CA 92069
(*Free Heat Machine®*)
Arizona Forest Supply, Inc.
610 E. Butler Ave.
Flagstaff, AZ 86002
Armstrong Products
P.O. Box 940
Huntington, WV 25701
Arnold Industries, Inc.
3011 Council St.
Toledo, OH 43606
Ashley Automatic Heater Co.
P.O. Box 1527
Huntsville, AL 35807
(*Wood burning heaters*)
Atlanta Stove Works, Inc.
P.O. Box 5254
Atlanta, GA 30307
Atlantic Clarion Stove Co.
Brewer, ME 04412
Autocrat Corp.
New Athens, IL 62264
Autostat
Howes Cave, NY 12092

Bahia Bar-B-Que
P.O. Box 1806
St. Paul, MN 55111
Bat Cave Stove Co.
Bat Cave, NC 28701

Bellway Mfg.
Grafton, VT 05146
Bestia Heater Ovens
Box 887
Charlestown, NH 03603
Better'n Ben's
Hayes-TE Equipment
P.O. Box 526Y
Unionville, CT 06085
Birmingham Stover Range
P.O. Box 2647
Birmingham, AL 35202
Black Bart Mfg.
5019 Hovis Rd.
Charlotte, NC 28208
Black Magic Chimney Sweeps
Box 977
Stowe, VT 05672
(*Chimney cleaning equipment*)
Black's
58 Maine St.
Brunswick, ME 04011
Blazing Showers
P.O. Box 327
Point Arena, CA 95468
Boston Stove Co.
155 John St.
Reading, MA 01867
Bow and Arrow Stove
14 Arrow St.
Cambridge, MA 02138
(*Godin stove*)
Buck Stove
116 Woodfin Ave.
Asheville, NC 28804
Buck Stoves of Columbus
P.O. Box 129
Westerville, OH 43081
Buffalo Stoves, International
P.O. Box 5526
Incline Village, WV 89450
Bullard Mfg. Co., Inc.
82 Kearney Ave.
Liverpool, NY 13088
The Burning Log
Box 438
Lebanon, NH 03766
Butler Stove Co.
Rt. 1 Box 32
Gooding, ID 83330

Canaqua Co.
1908 Baker St.
San Francisco, CA 94115
Carlson Mechanical Centers, Inc.
Box 242
Prentice, WI 54556
Carroll Mfg. Co.
P.O. Box 522 Eastwood Sta.
Syracuse, NY 13206
Cawley/LeMay Stove Co.
Box 431
Barto, PA 19504
Cawley Stove Co., Inc.
27 N. Washington St.
Boyertown, PA 19512
(*Airtight box stoves*)
Cedar Creek Homestead, Inc.
P.O. Box 666
Auburn, IN 46706
Ceramic Firepot
The Abercrombie Shop
P.O. Box 432Y
Oxford, MD 21654
Charmaster Prods., Inc.
2307 Highway 2 W.
Grand Rapids, MN 55744
Chim-A-Lator
8824 Wentworth Ave. S.
Minneapolis, MN 55420
Chopper Industires
P.O. Box 87
Easton, PA 18042
Columbia Prods. Co.
22 West Monroe St.
Chicago, IL 60603
Country Craftsmen
Box 3333-D
Santa Rosa, CA 95404
Country Stove
Box 460
Kent, CT 06757

Crescent Pipe and Nipple Co.
1950 Malila Rd.
Allegan, MI 49010
Cumberland Valley Metals, Inc.
P.O. Box 15666
Nashville, TN 37215

Daka Corp.
Industrial Pk.
Pine City, MN 55063
Dampney Co.
85 Paris St.
Everett, MA 02149
Damsite Dynamite Stove Co.
R.D. 3
Montpelier, VT 05602
Defiance Co.
Chassell, MI 49916
DeForge Industries
P.O. Box 216
Winooski, VT 05404
DeVault Fab-Weld & Piping Co.
Spring Mill Road
DeVault, PA 19432
Didier Mfg. Co.
1652 Phillips Ave.
Racine, WI 53403
Double Star
2466 Shattuck Ave.
Berkeley, CA 94704
Dover Corp.
P.O. Box 2015
Louisville, KY 40201
Dover Stove Co.
Main St.
Sangerville, ME 04479
Duel Fuel Prods., Inc.
2775 Pittsburgh Ave.
Cleveland, OH 44115

Earth Stove
19440 S.W. Boones Ferry Rd.
Tualatin, OR 97062
Edmund Scientific Co.
Barrington, NJ 08007
Ember Box Fireplaces
1012 S. Agnew
Oklahoma City, OK 73108
Empire Stove & Furnace Co.
793-7 Broadway
Albany, NY 12207
Energy Associates
P.O. Box 524
Old Saybrook, CT 06475
Energy Centers of America, Inc.
5276 Valley Industrial Blvd.
South Shakopee, MN 55379
Energy Harvesters Corp.
P.O. Box 19
Fitzwilliam, NH 03447
Energy Options, Inc.
P.O. Box 303
Green Bay, WI 54305

Fabsons Engineering
P.O. Box F-11
Leominster, MA 01453
Falcon Marketing Int.
P.O. Box 4567
Santa Rosa, CA 95401
Fireplace Center
2077 Jericho Turnpike
Commack, NY 11725
Fireplaces (N.S. Ltd.)
Suite 215, Duke St. Tower
Halifax, Nova Scotia B3J IN9
Fisher Prods., Inc.
Rt. 1, Box 63
Conifer, CO 80433
Fisher Stoves, Inc.
504 S. Main St.
Concord, NH 03301
Fisk Stove
Box 935
Dennis, MA 02638
Forest Fuels, Inc.
7 Main St.
Keene, NH 03431
Franco-Belge America
323 N. Lincoln
Niles, MI 49120

Franklin Cast Prods.
1800 Post Rd.
17 Airport Pl.
Warwick, RI 02886
Franklin Fireplaces
1100 Waterway Blvd.
Indianapolis, IN 46202
Free Flow Stove Wks.
Mine Rd.
S. Strafford, VT 05070
Fuego Heating Systems
P.O. Box 666
Brewer, ME 04412
Fuel Control Corp.
7661 Washington Ave. S.
Minneapolis, MN 55435
Futura Stove Works
Main St.
LaFarge, WI 54639

Garden Way Catalog
1300 Ethan Allen Ave.
Winooski, VT 05404
Garrison™ Stove Works, Inc.
Box 412
Claremont, NH 03743
Gemco
404 Main St.
Marlborough, NH 03455
General Products Corp.
150 Ardale St.
West Haven, CT 06516
Glenwood Stoves
R.D. 1
Ephrata, PA 17522
Glo-Fire
Spring & Summer Sts.
Lake Elsinore, CA 92330
Golden Enterprise
P.O. Box 422
Windsor, VT 05098
Goodtime Stove Co.
Rt. 9 Box 368
Williamsburg, MA 01096
Gorich Associates, Inc.
610 N. Cary St.
Brockton, MA 02402
Greenbriar Prods., Inc.
Box 473G
Spring Green, WI 53588
Green Mountain Stove Works, Inc.
Rt. 121 P.O. Box 477
Saxtons River, VT 05154
(*Barrell type stoves*)

Haagaard's Den
Bethel, ME 04217
Harbinger Prods.
Wheelock Rd.
Lyndonville, VT 05851
HDI Imports
Schoolhouse Farm
Etna, NH 03750
Hearth Craft
P.O. Box 20584
Portland, OR 97220
Heathdelle Sales Assoc., Inc.
Rt. 3
Meredith, NH 03253
Heatilator
Box 409
Mount Pleasant, IA 52641
Heat-N-Glo Fireplaces
1100 Riverwood Dr.
Burnsville, MN 55337
Heat Reclamation Div.
939 Chicopee St.
Chicopee, MA 01021
Heritage Fireplace Equipment Co.
1874 Englewood Ave.
Akron, OH 44312
Heritage Metalcraft Inc.
Rt. 202
S. Windham, ME 04082
Home Fireplaces
Markham, Ontario
Canada L3R IGE
Homestead Prods.
Star Route Box 275
Ramona, CA 92065

Household Woodsplitters
P.O. Box 143
Jeffersonville, VT 05464
Hunter Enterprises Orillia Ltd.
P.O. Box 400
Orilla, Ont. Canada
Hurricane Inc.
Rt. 5 Box 71A
Blackfoot, ID 83221
Hydraform Prods. Corp.
P.O. Box 2409
Rochester, NH 03867

Inglewood Stove Co.
Rt. 4
Woodstock, VT 05091
Integrated Thermal Systems
379 State St.
Portsmouth, NH 03801
Ironwood Stove Co.
115 N. First St.
Minneapolis, MN 55401
Island Stove Works, Inc.
894 Old Country Rd.
Westbury, NY 11590
Ispthermics, Inc.
P.O. Box 86
Augusta, NJ 07822

Jefferson Boat Works, Inc.
1655 N. Main St.
Jefferson, MA 01522
Jernlund Prods., Inc.
1620 Terrace Dr.
St. Paul, MN 55113
Johnson Energy Systems, Inc.
7350 N. 76th St.
Milwaukee, WI 53223

Kaufman Associates
234 Congress St.
Boston, MA 02110
(Thurmalox® 270 heat-resistant stove)
K&K Mfg., Inc.
552 Chestnut
Minonk, IL 61760
Kenematics
1140 N. Parker Dr.
Janesville, WI 53545
Kickapoo Stove Works Ltd.
Box 127
LaFarge, Wisconsin 54639
(*Kickapoo freestanding stove, furnace add-on, and woodburning home furnace*)
King Products Division
Martin Industries
P.O. Box 128
Florence, AL 35630
(*King automatic wood and coal circulators*).
Knotty Wood Splitters
Hebron, CT 06248
K.N.T. Inc.
P.O. Box 25
Hayesville, OH 44838
Kristia Assoc.
P.O. Box 1118
Portland, ME 04104

LaFont Corp.
1319 Town St.
Prentice, WI 54556
Lakewood Stoves
5608 Malvey RTSS
Fort Worth, TX 76107
Lance International
P.O. Box 562
Bloomfield, CT 06002
W.F. Landers Co.
P.O. Box 211
Springfield, MA 01101
Lassy Tools, Inc.
Plainville, CT 06062
Leyden Energy Conservation Corp.
Brattleboro Rd.
Leyden, MA 01337
Literlenot
P.O. Box 1881
Hickory, NC 28601
Locke Stove Co.
114 West 11 St.
Kansas City, MO 64105

Log House Designs
Fryeburg, ME 04058
Log Splitter
Thrust Mfg. Inc.
6901 S. Yosemite
Englewood, CO 80110
Longwood Furnace Corp.
Rt. 2 Box 223
Gallatin, MO 64640
Louisville Tint Stove Co.
P.O. Box 1079
Louisville, KY 40201
Lyndale Mfg. Co., Inc.
1309 N. Hills Blvd.
North Little Rock, AK 72116
Lyons Supply Co., Inc.
One Perimeter Rd.
Manchester, NH 03108

Maine Wood Heat Co.
Rt. 1 Box 38
Norridgewock, ME 04957
Majestic Co.
Huntington, IN 46750
Malleable Iron Range Co.
715 N. Spring St.
Beaver Dam, WI 53916
Malm Fireplaces, Inc.
368 Yolanda Ave.
Santa Rosa, CA 95404
Marathon Heater Co.
Box 165 R.D. 2
Marathon, NY 13803
Marco Industries, Inc.
P.O. Box 6
Harrisonburg, VA 22801
Markade-Winnwood
4200 Birmingham Rd.
Kansas City, MO 64117
Martin Industires
P.O. Box 730
Sheffield, AL 35660
Merry Music Box
20 Mckown
Boothbay Harbor, ME 04538
Metal Building Prods, Inc.
35 Progress Ave.
Hashua, NH 03060
Metal Concepts, Inc.
P.O. Box 25596
Seattle, WA 98125
Mitchell Creek Stove Co.
827 Parsons Rd.
Traverse City, MI 49684
Moderne-Aire
Highway 2 West
Grand Rapids, MI 55744
Mohawk Industries, Inc.
173 Howland Ave.
Adams, MA 01220
Monarch Range and Heater Div.
Malleable Iron Range Co.
715 North Spring St.
Beaver Dam, WI 53916
(*Add-a-furnace,® which is an add-on furnace for existing forced air heating systems*)

National Stove Works
Howe Caverns Road
Cobleskill, NY 12043
(*Thermo-control® woodburning system*)
New England Hardwood/Firewood Co.
Little City Road
Higganum, CT 06441
New Hampshire Stove Co.
R.F.D. 1
Brentwood, NH 03833
New Hampshire Wood Stoves, Inc.
Fairgrounds Rd.
Plymouth, NH 03264
Newmac Mfg. Inc.
Box 545
Woodstock, Ontario
Canada N4S 7W5
Nichols Environmental
5 Apple Rd.
Beverly, MA 01915
Nightingale Stove Co.
Handcrafted Woodburning Stoves
170 Old Main St.
Tewksbury, MA 01876

(*Octagonal-shaped freestanding wood stove*)
Nortech Corp.
300 Greenwood Ave.
Midland Park, NJ 07432

Old Country Appliances
P.O. Box 330-WD
Vacaville, CA 95688
Orley's Stoves
6999 6th St.
White City, OR 97501

Pioneer Energy Prods., Inc.
Rt. 1 Box 415
Forest, VA 24551
Pioneer Lamp & Stove Co.
105 South Washington,
Seattle, WA 98104
(*Pacific Princess iron cooking range*)
Pipe Prods. Co.
1274 Lincoln Rd.
Allegan, MI 49010
Portland Stove Foundry
57 Kennebec St.
Portland, ME 04104
Portland Willamette
6594 N.E. 59th Place
Portland, OR 97218
Preston Distributing Co.
10 Whidden St.
Lowell, MA 01852
Preway, Inc.
Wisconsin Rapids, WI 54494

Quaker Stove Company, Inc.
200 W. 5th St.
Lansdale, PA 19446
(*Moravian parlor stove*)

Radiant Grate
31 Morgan Park
Clinton, CT 06413
Rais & Wittus Inc.
Hack Green Road
Pound Ridge, NY 10576
(*Distributors of the Denmark wood stove*)

Ram and Forge
Brooks, ME 04921
REM Industries
408 C Simms Bldg.
Dayton, OH 45402
Ridgeway Steel Fabricators
Box 382
Ridgeway, PA 15853
Right Stove Polish Co., Inc.
P.O. Box 597W
Marlbourough, MA 01752
Riteway Manufacturing Co.
P.O. Box 153
Harrisonburg, VA 22801
(*Wood/coal heaters*)
RoKnich Products, Inc.
P.O. Box 311-E
No. Chicago, IL 60064
Russo Woodstove Mfg. Corp.
Water St.
Holbrook, MA 02343

Safe-T-Wedge Co.
1201 Candia Rd.
Manchester, NH 03103
Sally's Stoves
Box 174
Petaskala, OH 43062
Scandinavian Stoves, Inc.
Box 72
Alstead, NH 03602
Schrader Stoves, Inc.
660 Seneca
Eugene, OR 97402
Scot's Stove Co.
11 Ellis St.
Norwalk, CT 06850
Self Sufficiency Prods.
1 Appletree Sq.
Minneapolis, MN 55420
Shayer Stove Co., Inc.
34 C Street NE
Auburn, WA 98002
Shenandoah Mfg. Co., Inc.
P.O. Box 839
Harrisonburg, VA 22801
Solar Sauna
Box 466
Hollis, NH 03049

Solis Energy Works
Box 262A
New Hope, PA 13938
Sotz Corp.
23797 Sprague Rd.
Columbia Station, OH 44028
Southeastern Vermont Comm.
Action, Inc.
7-9 Westminster St.
Bellows Falls, VT 05701
Southport Stoves
248 Tolland St.
East Hartford, CT 06108
Southwest Mfg.
10 N. Elliot St.
Aurora, MO 65605
Spencer Prods. Co.
1134 Poplar Place S.
Seattle, WA 98144
STIHL, Inc.
P.O. Box 5514
Virginia Beach, VA 23455
Stove Works
P.O. Box 172A
Marlboro, VT 05344
Sturges Heat Recovery, Inc.
P.O. Box 397
Stone Ridge, NY 12484
Suburban Mfg. Co.
P.O. Box 6472
Columbia, SC 29206
Sunrise Energy Prods.
825 Charlevoix Ave.
Petoskey, MI 49770
Sunshine Stove Works®, Inc.
Brooks, ME 04921
(Sunshine stove based on a scandinavian system)
Superior Fireplace Co.
P.O. Box 2066
Fullerton, CA 92633
Svendborg Co., Inc.
P.O. Box 5
Hanover, NH 03755

Tekton Design Corp.
Conway, MA 01341
Temco
P.O. Box 1184
Nashville, TN 37202
Thermo Control Wood Stoves
Central Bridge, NY 12035
Thermograte, Inc.
P.O. Box 43566
St. Paul, MN 55164
(Thermograte®, a freestanding fireplace store or an insert for fireplaces)
Tilton Equipment Co.
4612 N. Chatsworth
St. Paul, MN 55112
Timberline Stoves Int. Inc.
5231 Chirden Blvd.
Boise, ID 83704
Tirolia of America, Inc.
169 Dunning Rd.
Middletown, NY 10950
Top O' the Stove Prods.
Ferry Rd.
Grand Isle, VT 05458
Torrid Air®
Torrid Manufacturing Co., Inc.
1248 Poplar Pl. So.
Seattle, WA 98144
(Torrid Air® stove)
Townsend Chain Saw Co.
P.O. Box 6396
Columbia, SC 29260
Triway Mfg. Inc.
Box 37
Marysville, WA 98270

United States Stove Co.
P.O. Box 151
South Pittsburgh, TN 37380
Upland Stove Co., Inc.
P.O. Box 87
Greene, NY 13778

Valco Corp.
215 Johnson Rd.
Michigan City, IN 46360
Valley Forge Stove Co.
85 East Bridge St.
Spring City, PA 19475

Vaporpack, Inc.
Box 428
Exeter, NH 03833
Vermont Counterflow Wood Furnace
Plainfield, VT 05667
Vermont Soapstone Co.
Perkinsville, VT 05151
Vermont Woodstove Co.
P.O. Box 1016
Bennington, VT 05201
Volunteers in Tech. Asst.
3706 Rhode Island Ave.
Mt. Ranier, MD 20822

Warner Stove
P.O. Box 292
Warner, NH 03278
E.G. Washburne and Co.
83 Andover St.
Danvers, MA 01923
Washington Stove Works
P.O. Box 687
Everett, WA 98201
Weathercock Industries, Inc.
1963 West Sixth
Eugene, OR 97402
White Mesa, Inc.
110 Laguna
Albuquerque, NM 87104
Whittier Steel & Mfg. Inc.
10725 S. Painter Ave.
Santa Fe Springs, CA 90670
Whole Earth Access Co.
2466 Shattuck Ave.
Berkeley, CA 94702
Wilkeining Mfg. Co.
Walker, MN 56484
Will-Burt Co.
169 S. Main
Orrville, OH 44667
Wilson Industries
2296 Wycliff
St. Paul, MN 55114
Woodburning Specialties
P.O. Box 5
No. Marshfield, MA 01059
Wood Heat, Inc.
1748 Warm Springs Rd.
Columbus, GA 31904
Woodking Stoves International of Canada, Inc.
1145 Fife St.
Winnepeg, Manitoba
Canada R2X 2N2
Woodstock Soapstone Co.
Bridgewater Mill Mall
Bridgewater, VT 05034
Worcester Brush Co.
Box 658-WD
Worcester, MA 01601

Yankee Woodstoves
P.O. Box 7
Bennington, NH 03442

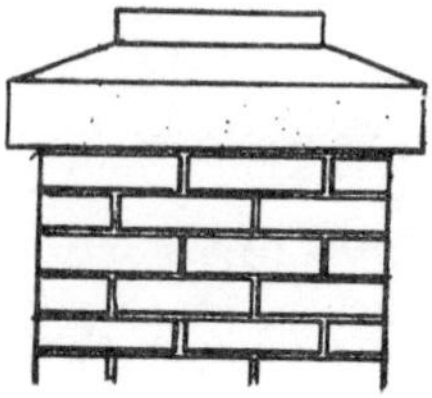

Index